Basics of
ABSTRACT ALGEBRA

Basics of ABSTRACT ALGEBRA

Rohit Garg

ANMOL PUBLICATIONS PVT. LTD.
NEW DELHI-110 002 (INDIA)

ANMOL PUBLICATIONS PVT. LTD.
Regd. Office: 4360/4, Ansari Road, Daryaganj,
New Delhi-110002 (India)
Tel.: 23278000, 23261597, 23286875, 23255577
Fax: 91-11-23280289
Email: anmolpub@gmail.com
Visit us at: www.anmolpublications.com

Branch Office: No. 1015, Ist Main Road, BSK IIIrd Stage
IIIrd Phase, IIIrd Block, Bengaluru-560 085 (India)
Tel.: 080-41723429 • Fax: 080-26723604
Email: anmolpublicationsbangalore@gmail.com

Basics of Abstract Algebra

First Edition, 2011

ISBN 978-81-261-4899-8

PRINTED IN INDIA

Printed at Suman Printers, Delhi

Contents

Preface

Abstract algebra is the subject area of mathematics that studies algebraic structures, such as groups, rings, fields, modules, vector spaces, and algebras. The phrase abstract algebra was coined at the turn of the 20th century to distinguish this area from what was normally referred to as algebra, the study of the rules for manipulating formulae and algebraic expressions involving unknowns and real or complex numbers, often now called elementary algebra. The distinction is rarely made in more recent writings.

Contemporary mathematics and mathematical physics make extensive use of abstract algebra; for example, theoretical physics draws on Lie algebras. Subject areas such as algebraic number theory, algebraic topology, and algebraic geometry apply algebraic methods to other areas of mathematics. Representation theory, roughly speaking, takes the 'abstract' out of 'abstract algebra', studying the concrete side of a given structure; see model theory.

Author

Preface

1

Vector Space

SCALAR FIELD

Vector calculus is concerned with scalar fields, which associate a scalar to every point in space, and vector fields, which associate a vector to every point in space.

For example, the temperature of a swimming pool is a scalar field: to each point we associate a scalar value of temperature.

The water flow in the same pool is a vector field: to each point we associate a velocity vector.

Vector calculus is a field of mathematics concerned with multivariable real analysis of vectors in an inner product space of two or more dimensions (some results — those that involve the cross product — can only be applied to three dimensions).

It consists of a suite of formulae and problem solving techniques very useful for engineering and physics.

VECTORSH

We have three types of physical quantities:

1. *Scalars:* Magnitude Only. *e.g.* Mass, Length and Time.
2. *Vectors:* Magnitude and Direction. *e.g.* Velocity and Force.
3. *Tensors:* Not covered here.

The simplest way to differentiate vectors from scalars is to take Speed vs. Velocity.In the path shown above, assume that the particle has a fixed speed. While it's speed does not change, it does have an acceleration, as the direction changes constanly.

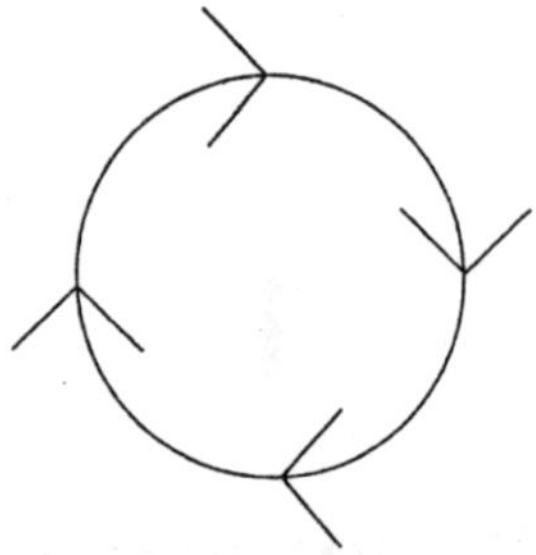

Free Vectors vs. Bound Vectors

In this course we will be dealing with free vectors. Free vectors are deemed to be equal if they have the same magnitude and direction, regardless of their location. The same does not hold true for bound vectors.

Free Vectors

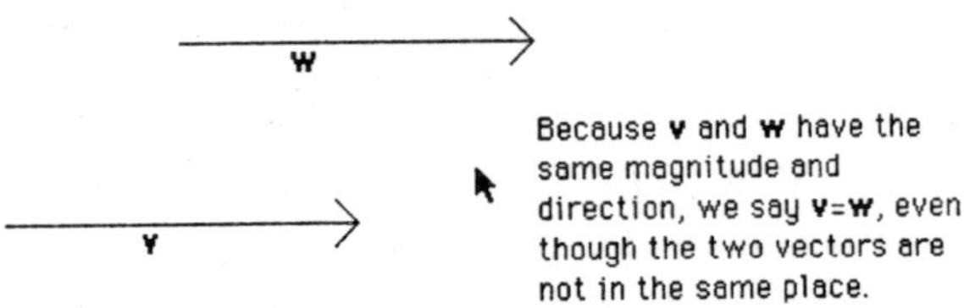

Bound Vectors

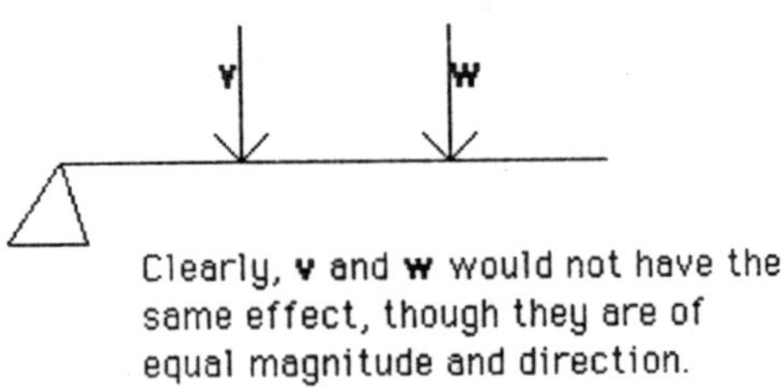

Unit Vectors

- *Def*: A unit vector is a vector of length (magnitude) one. The 2–D space with which we are very familiar has two particular unit vectors reffered to as a "Basis." They are the i and j vectors. Because they are a Basis every vector can be written in terms of these two vectors. For example $v = 3i + 2j$. Thus, one can determine the components of v. It is important; however, to differentiate between the scalar and vector components.

- *Scalar Components*: 3,2 Vector Components: $3i$, $2j$

Laws of Addition

There are two ways to add vectors, geometrically and algebraically.

Geometric Addition

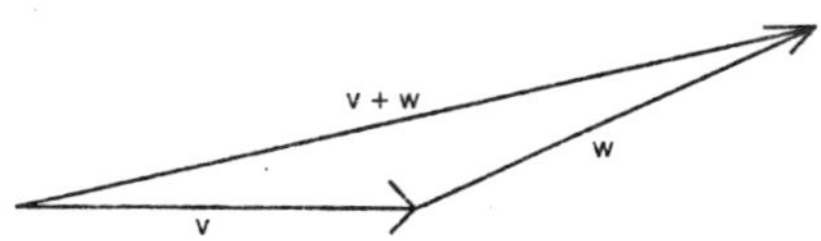

The parrallelogram method of vector addition

Algebraic Addition

$$v = 3i - 2j,\ w = i + j$$
$$v + w = 4i - j$$

Magnitude

The Magnitude of a vector is it's length, defined in 2-D as below.

Magnitude of a vector v:

$$|v| = |ai + bj| = +\sqrt{a^2 + b^2}$$

Direction

Def: The direction of v is a unit vector parallel to v

$$\text{dir } v = \frac{v}{|v|}.$$
$$v = 2i + 2j$$

$$|v| = 5$$

$$\text{dir } v = \frac{3}{5}i + \frac{4}{5}j$$

For Example: But what does this mean? We usually think of direction in terms of angles, not components of vectors. This brings us to slope.

Slope

To find θ requires using the double argument Arctangent. Since information like slope = 1 would yield both 45° and 135° for θ.

$$v = ai + bj$$

$$r = \sqrt{a^2 + b^2}$$

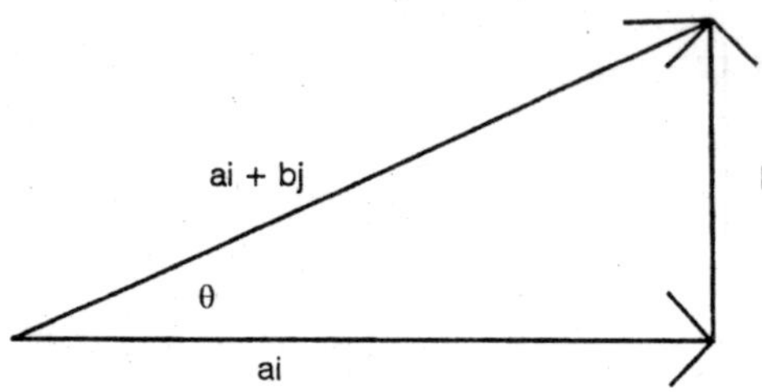

Then Slope:

$$a \qquad = r * \cos\theta$$

$$\text{slope} = b/_a = \tan\theta \; a \qquad = r * \sin\theta$$

This can be used to find unit tangents and normals to curves defined in the x–y plane. Any curve will have two unit tangents and two unit normals at any point, so it is necessary to differentiate betweent the clockwise and counterclockwise tangents and the in and out unit normals. The example below shows an inward unit normal and a clockwise unit tangent.

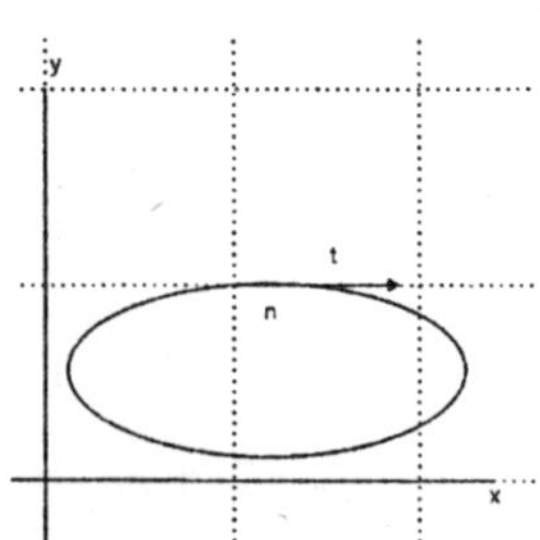

In order to find these tangents, it is necessary to "marry" calculus and slope. Since the derivitive, dy/dx, is equal to the slope, b/a, you can find the unit tangent vector simply by setting these two quantities equal to each other.

Namely:

$$dy/dx = b/a$$

GEOMETRY OF VECTORS

Last lecture left off discussing the "marriage" of vectors and calculus in order to find a vector tangent to a given curve in the x–y plane. The following example illustarates exactly how this can be done.

The circle is $x^2 + y^2 = 1$.

Point P has coordinate $(1/\sqrt{2}, 1/\sqrt{2})$

We want the counterclockwise tangent and outward normal pictured above

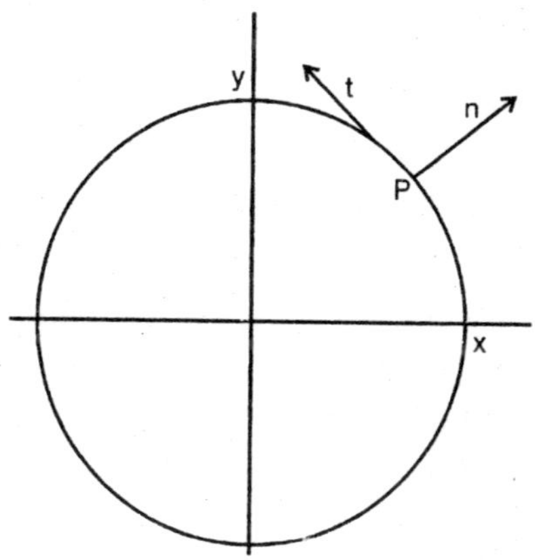

First find the slope of the tangent: $2x + 2y\dfrac{dy}{dx}$

$$= 0\frac{dy}{dx} = -\frac{x}{y} = -1$$

Also. slope = b/a

So $b/a = -1$ and $b = -a$

Therefore $T = i - j$ or $T = j - i$

Looking at the picture we see t up and left. or negative x and positive y,

so $T = j - i$. Normalizing (getting the unit vector), t

$$= \frac{1}{\sqrt{2}}(j - i).$$

For the Normal: Slope of the normal is the neg. reciprocal of the tangent (They're ⊥.) = 1. So $N = i + j$ or $N = -i - j$. Since the normal has pos. x and y components, we know that $N = i + j$, normallizing $n = \frac{1}{\sqrt{2}}(i - j)$

3-D Cartesian Coordinates

In three dimensions, the two dimentional concepts are extended, with the simple addition of a third component-the unit vector parallel to the z axis, k. Magnitude and direction are still computed in extremely similar ways.

$$v = ai + bj + ck$$

$$|v| = +\sqrt{a^2 + b^2 + c^2}$$

$$\text{dir}\, v = \frac{v}{|v|}$$

$$\text{Ex: } v = i - 2j + k$$

$$\text{dir}\, v = \frac{1}{\sqrt{6}}(i - 2j + k)$$

We would still like to have angles to describe direction rather than a difficult to grasp set of components. Two Dimentions required one angle to define a direction. In three dimensions we'll need two.

Spherical Coordinates

Spherical coordinates define a vector in terms of its length (R), an angle between the vector and the z-axis (phi), and the angle between the vectors projection onto the xy plane and the x-axis (theta).

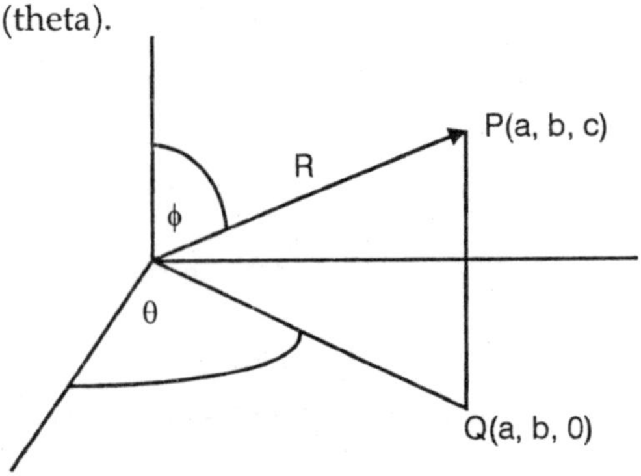

By looking at the right triangles formed by the vector, its projection, and the axes, the relationship between the Spherical coordinates and Cartesian ones can be determined. It turns out that $x = R$ sin (phi) cos (theta); $y = R$ sin (phi) sin (theta); and $z = R$ cos (phi).

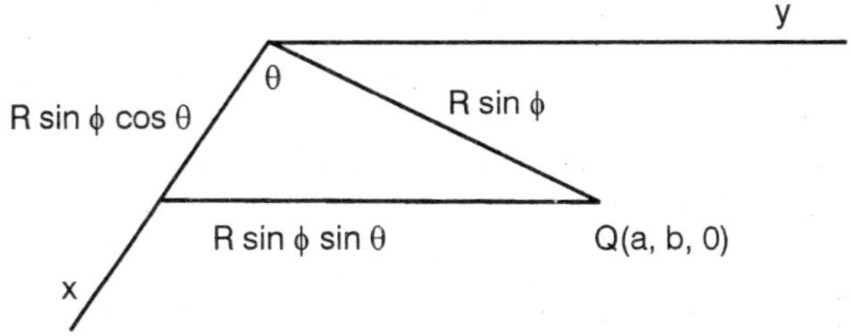

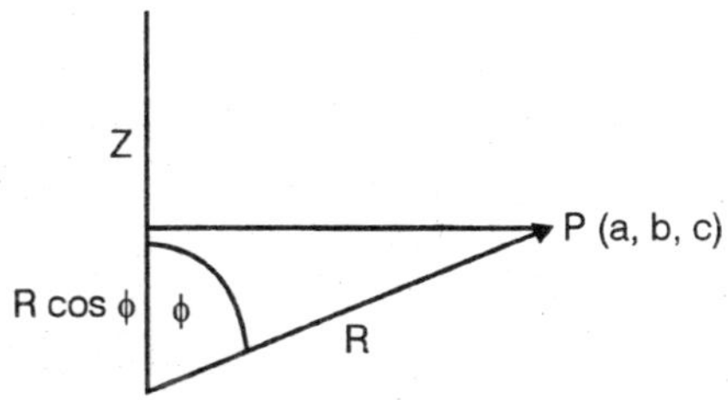

3D Geometry in three Dimentions there are four types of objects that can Appear

1. *Surfaces:* Surfaces are of the form $f(x, y, z) = 0$ and are things like spherical shells or planes. *Ex:* $x^*x + y^*y + z^*z = 1$ [sphere] or $ax + by + cz = d$ [plane].
2. *Lines:* A line is the intersection of two surfaces and is thus defined by a system of two equations, namely, $f1(x, y, z) = 0$; $f2(x, y, z) = 0$ *Ex*: $x = 2$; $y = 3$.
3. *Volume:* Volumes are of the form $f(x, y, z) << 0$. Examples include things like a solid sphere $x^*x + y^*y + z^*z << 1$.
4. *Point:* (1,2,3).

FUNCTIONS OF SEVERAL VARIABLES

Suppose we go beyond $f(x)$ and talk about $F(x,y,z)$–*e.g.* a function of the *exact position in space.* This is just an example, of course; the abstract idea of a function of several variables can have "several" be as many as you like and "variables" be anything you choose. Another place where we encounter lots of functions of "several" variables is in THERMODYNAMICS, but for the time being we will focus our attention on the three *spatial* variables x (left-right), y (back-forth) and z (up-down).

How can we tackle *derivatives* of this function?

PARTIAL DERIVATIVES

Well, we do the obvious: we say, "Hold all the *other* variables *fixed* except [for instance] x and then treat $F(x,y,z)$ as a function only of x, with y and z as fixed *parameters.*" Then we know just how to define the derivative with respect to x. The short name for this derivative is the PARTIAL DERIVATIVE *with respect to x,* written symbolically

$$\frac{\partial F}{\partial x}$$

where the fact that there are other variables being held fixed is implied by the use of the symbol ∂ instead of just d.

Similarly $\frac{\partial F}{\partial y}$ for and $\frac{\partial F}{\partial z}$.

OPERATORS

The foregoing description applies for *any* function of (x,y,z); the concept of "taking partial derivatives" is independent of what function we are taking the derivatives *of*. It is therefore practical to learn to think of,

$$\frac{\partial}{\partial x} \text{ and } \frac{\partial}{\partial y} \text{ and } \frac{\partial}{\partial y}$$

as Operators that can be applied to *any* function (like F). Put the operator on the left of a function, perform the operation and you get a partial derivative–a new function of (x,y,z). In general, such "operators" *change one function into another.* Physics is loaded with operators like these.

THE GRADIENT OPERATOR

The GRADIENT operator is a *vector* operator, written $\vec{\nabla}$ and called "grad" or "del." It is defined (in Cartesian coordinates x,y,z) as[1].

$$\vec{\nabla} = \hat{i}\frac{\partial}{\partial x} + \hat{j}\frac{\partial}{\partial y}\hat{k}\frac{\partial}{\partial z}$$

and can be applied directly to any *scalar* function of (x,y,z)– say, $\phi(x,y,z)$ to turn it into a *vector* function,

$$\vec{\nabla}\phi = \hat{i}\frac{\partial}{\partial x} + \hat{j}\frac{\partial}{\partial y}\hat{k}\frac{\partial}{\partial z}.$$

GRADIENTS OF SCALAR FUNCTIONS

It is instructive to work up to this "one dimension at a time." For simplicity we will stick to using f as the symbol for the function of which we are taking derivatives.

The GRADIENT in One Dimension

Let the dimension be x. Then we have no "extra" variables to hold constant and the gradient of f (x) is nothing but,

$$\hat{i}\frac{d\phi}{\partial x}.$$

We can illustrate the "meaning" of $\vec{\nabla}$f by an example: let f (x) be the mass of an object times the acceleration of gravity times the height h of a hill at horizontal position x. That is f (x), is the *gravitational potential energy* of the object when it is at horizontal position x.

Then,

$$\vec{\nabla}\phi = \hat{i}\frac{d\phi}{\partial x} + \hat{i}\frac{\partial}{\partial x}(mgh) = mg\left(\frac{\partial h}{\partial x}\right)i.$$

Note that,

$$\frac{\partial h}{\partial x}$$

is the *slope* of the hill and$-\vec{\nabla}\phi$ is the *horizontal component of the net force* (gravity plus the normal force from the hill's surface) on the object. That is, $-\vec{\nabla}\phi$ is the *downhill force.*

The GRADIENT in Two Dimensions

In the previous example we disregarded the fact that most hills extend in *two* horizontal directions, say x = East and y = North. In this case there are two *components* to the slope: the

Eastward slope,

$$\frac{\partial h}{\partial x}$$

and the Northward slope,

$$\frac{\partial h}{\partial y}.$$

The former is a measure of how steep the hill will seem if you head due East and the latter is a measure of how steep it will seem if you head due North. If you put these together to form a *vector* "steepness".

$$\vec{\nabla} h = \hat{i}\frac{\partial h}{\partial x} + \hat{j}\frac{\partial h}{\partial y}$$

then the vector $\vec{\nabla} h$ points *uphill–i.e.* in the direction of the *steepest ascent*. Moreover, the gravitational potential energy $\phi = mgh$ as before [only now is a function of 2 variables, $\phi(x, y)$] so that $-\vec{\nabla}\phi$ is once again the *downhill force* on the object.

The Gradient in Three Dimensions

If the potential ϕ is a function of 3 variables, $\phi(x, y, z)$ [such as the three *spatial coordinates* x, y and z–in which case we can write it a little more compactly as $\phi(\vec{r})$ where $\vec{r} \equiv x\hat{i} + y\hat{j} + z\hat{k}$, the vector distance from the origin of our coordinate system to the point in space where ϕ is being evaluated], then it is a little more difficult to make up a "hill" analogy—try imagining a topographical map in the form of a 3-dimensional hologram where instead of *lines* of constant *altitude* the "equipotentials" are *surfaces* of constant ϕ. In general, any time there is a *potential energy* function $\phi(\vec{r})$ we can immediately write down the *force* $\vec{F}$ associated with it as

$$\vec{F} \equiv -\vec{\nabla}\phi$$

A perfectly analogous expression holds for the *electric field* $\vec{E}$ [force per unit charge] in terms of the *electrostatic potential* ϕ [potential energy per unit charge]:[2]

$$\vec{E} \equiv -\vec{\nabla}\phi$$

The GRADIENT in *N* Dimensions

Although we won't be needing to go beyond 3 dimensions

very often in Physics, you might want to borrow this metaphor for application in other realms of human endeavour where there are more than 3 variables of which your scalar field is a function.

You could have f be a measure of *happiness*, for instance; then f might be a function of lots of factors, such as x_1 = freedom from violence, x_2 = freedom from hunger, x_3 = freedom from poverty, x_4 = freedom from oppression, and so on.[3] Note that with an arbitrary number of variables we get away from thinking up different names for each one and just call the ith variable "x_i."

Then we can define the gradient in N dimensions as,

$$\vec{\nabla}\phi = \hat{i}_1 \frac{\partial\phi}{\partial x_1} + \hat{i}_2 \frac{\partial\phi}{\partial x_2} + \ldots + \hat{i}_N \frac{\partial\phi}{\partial x_N}$$

or

$$\vec{\nabla}\phi = \sum_{i=1}^{N} \hat{i}_i \frac{\partial\phi}{\partial x_i}$$

where $\hat{i}_i$ is a unit vector in the x_i direction.

DIVERGENCE OF A VECTOR FIELD

If we form the scalar ("dot") product of $\vec{\nabla}$ with a *vector* function $\vec{A}(x,y,z)$ we get a *scalar* result called the DIVERGENCE of $\vec{A}$:

$$\text{div}\,\vec{A} \equiv \vec{\nabla} \cdot \vec{A} \equiv \frac{\partial A_x}{\partial x} + \frac{\partial A_y}{\partial y} + \frac{\partial A_z}{\partial z}$$

This name is actually quite mnemonic: the divergence of a vector field is a *local* measure of its "outgoingness"—*i.e.* the extent to which there is more *exiting* an infinitesimal region of space than *entering* it.

If the field is represented as "flux lines" of some indestructible "stuff" being emitted by "sources" and absorbed by "sinks," then a nonzero divergence at some point means there must be a *source* or *sink* at that position.

That is to say, *"What leaves a region is no longer in it."*

For example, consider the divergence of the current density $\vec{J}$, which describes the flux of a conserved quantity such as electric charge Q.

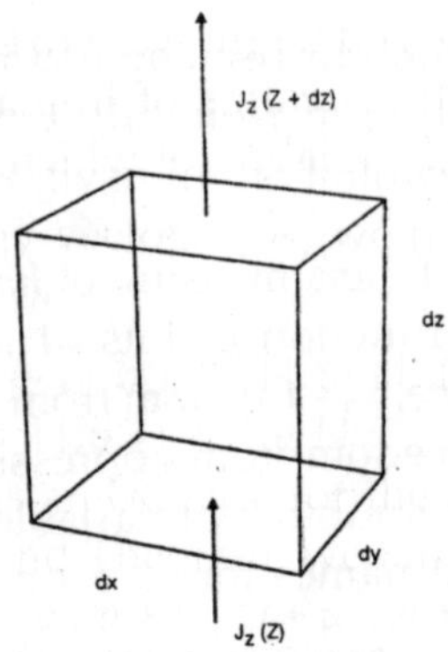

Fig. Flux into and out of a Volume Element.

$$dV = dx\ dy\ dz.$$

To make this as easy as possible, let's picture a *cubical* volume element $dV = dx\ dy\ dz$. In general, $\vec{J}$ will (like any vector) have three components (J_x, J_y, J_z), each of which may be a function of position (x,y,z). If we take the lower left front corner of the cube to have coordinates (x,y,z) then the upper right back corner has coordinates $(x + dx, y + dy, z + dz)$. Let's concentrate first on J_z and how it depends on z.

It may not depend on z at all, of course. In this case, the amount of Q coming into the cube through the bottom surface (per unit time) will be the same as the amount of Q going out through the top surface and there will be no net gain or loss of Q in the volume-at least not due to J_z.

If J_z is bigger at the top, however, there will be a net loss of Q within the volume dV due to the "divergence" of J_z. Let's see how much: the difference between $J_z(z)$ at the bottom and $J_z(z + dz)$ at the top is, by definition, $dJ_z = \left(\frac{\partial J_z}{\partial z}\right) dz$. The flux is over the same area at top and bottom, namely $dx\ dy$, so the total *rate of loss* of Q due to the z-dependence of J_z is given by

$$\dot{Q}_z = -dx\,dy\left(\frac{\partial J_z}{\partial z}\right)dz = -\left(\frac{\partial J_z}{\partial z}\right)dx\,dy\,dz$$

or

$$\dot{Q} = -\left(\frac{\partial J_z}{\partial z}\right)dV.$$

A perfectly analogous argument holds for the x-dependence if J_x and the y-dependence of J_y, giving a total rate of change of Q

$$\dot{Q} = \left(\frac{\partial J_x}{\partial x} + \frac{\partial J_y}{\partial y} + \frac{\partial J_z}{\partial z}\right) dV$$

or

$$\dot{Q} = -\vec{\nabla} \cdot \vec{J}\, dV$$

The total amount of Q *in* our volume element dV at a given instant is just ρdV, of course, so the rate of change of the enclosed Q is just

$$\dot{Q} = \dot{\rho}\, dV$$

which means that we can write

$$@ -\vec{\nabla} \cdot \vec{J}\, dV$$

or, just cancelling out the common factor dV on both sides of the equation,

$$\frac{\partial \rho}{\partial t} = -\vec{\nabla} \cdot \vec{J}$$

which is the compact and elegant "differential form" of the equation of continuity.

This equation tells us that the "Q sourciness" of *each point* in space is given by the degree to which flux "lines" of $\vec{J}$ tend to radiate *away* from that point more than they converge *toward* that point–namely, the divergence of $\vec{J}$ at the point in question. This esoteric-looking mathematical expression is, remember, just a formal way of expressing our original dumb tautology!

CURL OF A VECTOR FIELD

If we form the vector ("cross") product $\vec{\nabla}$ of with a *vector* function $\vec{A}(x, y, z)$ we get a *vector* result called the curl of $\vec{A}$:

$$\text{curl}\, \vec{A} \equiv \vec{\nabla} \times \vec{A}$$

$$\equiv \hat{i}\left(\frac{\partial A_z}{\partial y} - \frac{\partial A_y}{\partial z}\right) + \hat{j}\left(\frac{\partial A_x}{\partial z} - \frac{\partial A_z}{\partial x}\right) + \hat{k}\left(\frac{\partial A_y}{\partial x} - \frac{\partial A_x}{\partial y}\right)$$

This is a lot harder to visualize than the DIVERGENCE,

but not impossible. Suppose you are in a boat in a huge river (or Pass) where the current flows mainly in the x direction but where the speed of the current (flux of water) varies with y. Then if we call the current $\vec{J}$, we have a nonzero value for the derivative $\frac{\partial A_x}{\partial y}$, which you will recognize as one of the terms in the formula for. What does this imply? Well, if you are sitting in the boat, moving with the current, it means the current on your port side moves faster–*i.e.* forward relative to the boat–and the current on your starboard side moves slower–*i.e.* backward relative to the boat–and this implies a *circulation* of the water around the boat–*i.e.* a *whirlpool!* So $\vec{\nabla} \times \vec{J}$ is a measure of the local "swirliness" of the current $\vec{J}$, which means "curl" is not a bad name after all!

THE LAPLACIAN OPERATOR

If we form the scalar ("dot") product of $\vec{\nabla}$ with *itself* we get a scalar *second derivative* operator called the laplacian:

$$\vec{\nabla} \cdot \vec{\nabla} \equiv \nabla^2 = \frac{\partial^2}{\partial x^2} + \frac{\partial^2}{\partial y^2} + \frac{\partial^2}{\partial z^2}$$

What does the ∇^2 operator "mean?" It is the three-dimensional generalization of the one-dimensional CURVATURE operator d^2/dx^2. Consider the familiar one-dimensional function $h(x)$ where h is the height of a hill at horizontal position x. Then dh/dx is the *slope* of the hill and d^2h/dx^2 is its *curvature* (the rate of change of the slope with position). This property appears in every form of the WAVE EQUATION. In three dimensions, a nice visualization is harder (there is no extra dimension "into which to curve") but $\nabla^2\phi$ represents the equivalent property of a scalar function $\phi(x, y, z)$.

GEOMETRY IN HIGHER DIMENSIONS

THE CROSS PRODUCT

We define the cross product of two three-dimensional vectors a and b by the requirements:

1. $a \times b$ is perpendicular to both a and b.

2. $||a \times b||$ is the area of the parallelogram spanned by a and b (*i.e.* the parallelogram whose adjacent sides are the vectors a and b).
3. The direction of $a \times b$ is determined by right-hand rule. (This means that if we curl the fingers of the right hand from a to b, then the thumb points in the direction of $a \times b$.)

Recall from lecture (or your textbook) that the area of the parallelogram is $||a|| \; ||b||$ sin, where is the angle between a and b.

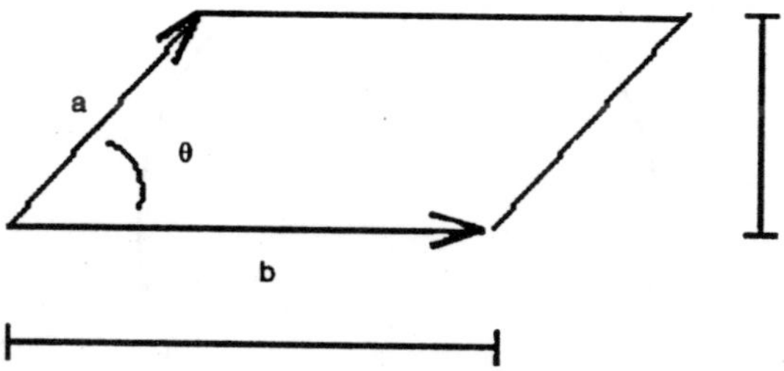

This means that the magnitude of the cross product is largest when a and b are perpendicular.

On the other hand, if a and b are parallel or if either vector is the zero vector, then the cross product is the zero vector.

The Component Formula for the cross Product

Recall the geometric definition of the cross product as given in an earlier reading.

Cross Product of Unit Vectors

First, we derive the cross product of the unit vectors, i, j, and k be the standard unit vectors in R^3. (We define the cross product only in three dimensions.)

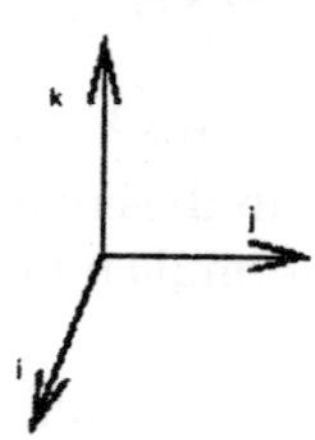

The parallelogram spanned by any two of these standard

unit vectors is a unit square, which has area one. Hence, the cross product must be a unit vector. Looking at the above graph, you can use the right-hand rule to determine the following results.

$$i \times j = k$$
$$j \times k = i$$
$$k \times i = j$$

This little cycle diagram can help you remember these results.

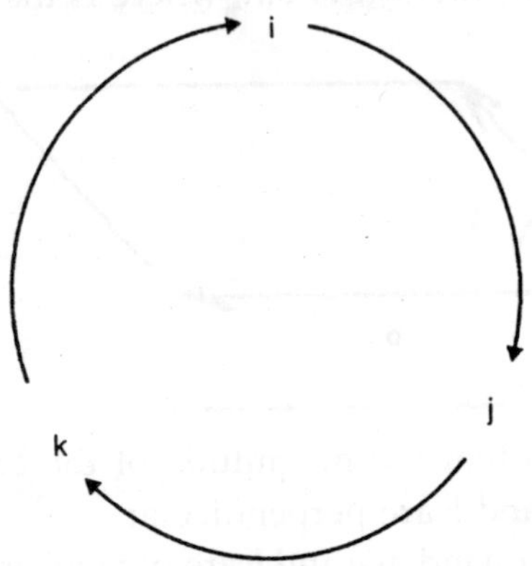

What about $i \times k$? By the right-hand rule, it must be$-j$. I think you get the idea.

General Vectors

In an earlier reading, we mentioned the special properties of the cross product: $b \times a = -a \times b$, and $a \times a = 0$. With these exceptions, the cross product behaves like regular multiplication, obeying these properties,

1. $(ya) \times b = y(a \times b) = a \times (yb)$,
2. $a \times (b + c) = a \times b + a \times c$,
3. $(b + c) \times a = b \times a + c \times a$,

where a, b, and c are vectors in R^3 and y is a scalar. We can use these properties, along with the cross product of the standard unit vectors, to write the formula for the cross product in terms of components.

We write the Components of a and b as:

$$a = (a_1, a_2, a_3) = a_1 i + a_2 j + a_3 k$$

$$b = (b_1, b_2, b_3) = b_1 i + b_2 j + b_3 k$$

First, we'll assume that $a_3 = b_3 = 0$. (Then, the manipulations are much easier.) We calculate:

$$\begin{aligned} a \times b &= (a_1 i + a_2 j) \times (b_1 i + b_2 j) \\ &= a_1 b_1 (i \times i) + a_1 b_2 (i \times j) + a_2 b_1 (j \times i) + a_2 b_2 (j \times j) \end{aligned}$$

Since we know that $i \times i = 0 = j \times j$ and that $i \times j = k = -j \times i$, this quickly simplifies to

$$\begin{aligned} a \times b &= (a_1 b_2 - a_2 b_1) k \\ &= \begin{vmatrix} a_1 & a_2 \\ b_1 & b_2 \end{vmatrix} k. \end{aligned}$$

Writing the result as a determinant, as we did in the last step, is a handy way to remember the result. What about the general case where a_3 and b_3 might not be zero? Well, that's too many manipulations for me to write it out here, and you probably wouldn't read it anyway. To calculate the result, it's just the same process as above. You can verify yourself that in general

$$\begin{aligned} a \times b &= (a_1 i + a_2 j + a_3 k) \times (b_1 i + b_2 j + b_3 k) \\ &= \begin{vmatrix} a_2 & a_3 \\ b_2 & b_3 \end{vmatrix} i - \begin{vmatrix} a_1 & a_3 \\ b_1 & b_3 \end{vmatrix} j + \begin{vmatrix} a_1 & a_2 \\ b_1 & b_2 \end{vmatrix} k. \end{aligned}$$

Looking at the formula for the 3×3 determinant, we see that the formula for a cross product looks a lot like the formula for the 3×3 determinant.

If we allow a matrix to have the vector i, j, and k as entries, the 3×3 determinant gives a handy mnemonic to remember the cross product:

$$a \times b = \begin{vmatrix} i & j & k \\ a_1 & a_2 & a_3 \\ b_1 & b_2 & b_3 \end{vmatrix}$$

This is the way I remember how to compute the cross product.

COORDINATES CARTESIAN COORDINATE SYSTEM IN THE PLANE AND IN SPACE

A Cartesian coordinate system in the space consists of

number lines (axes) perpendicularly intersecting in one point O called the origin. The orientation of the axes is chosen according to the„Right-Hand-Rule“ as shown in the figure.

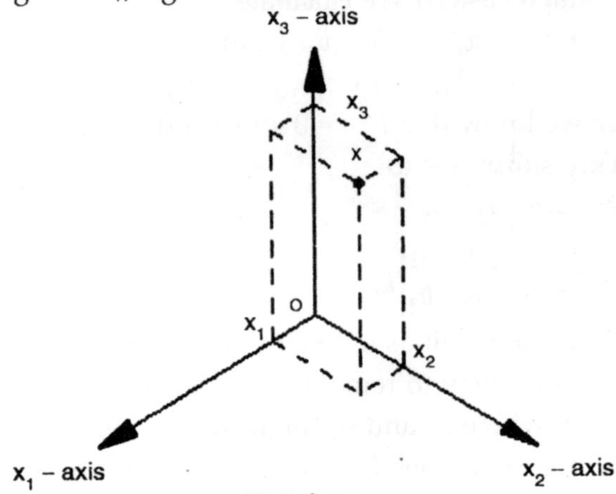

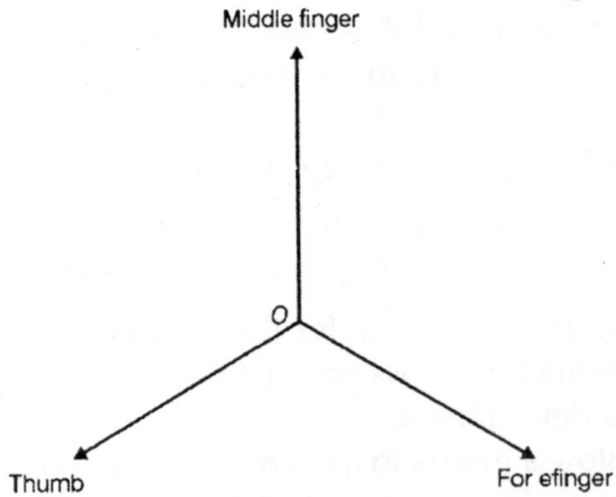

A point X is determined by the values x_i-called the coordinates-obtained by perpendicular projection of the point X onto the axes: $X = (x_1, x_2, x_3)$. Using a notation without subscripts, the coordinates are usually symbolized by (x, y, z) and the number axes are called x –, y–and z-axis. A Cartesian coordinate system in the plane is defined in an analogous manner.

PIXEL COORDINATES

When drawing to a screen, the depicted objects are often given by pixel coordinates. The depicted rectangle corresponds to a window with 90 × 120 pixels and the coordinates of the coloured pixels refer to the upper left corner of the window.

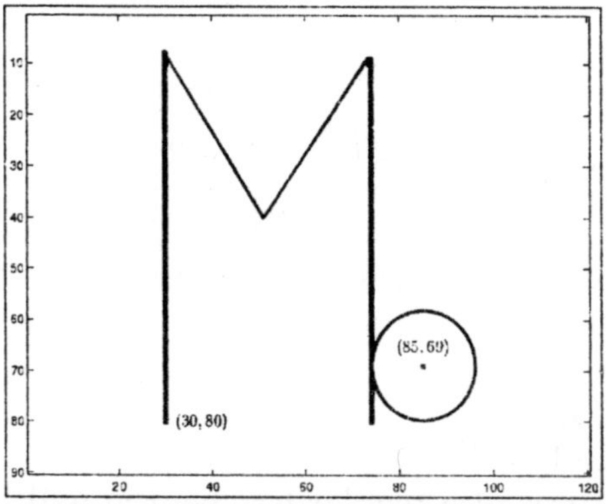

SPHERICAL COORDINATES

A point $P = (x, y, z)$ can be represented by the following 3 values: the point's distance from the origin $r = |\overline{OP}|$, the angle $\varphi \in [0, 2\pi)$ formed by the x-axis and the projection of $\overline{OP}$ onto the xy-plane, and the angle $\vartheta \in [0, \pi]$ formed by and the z-axis.

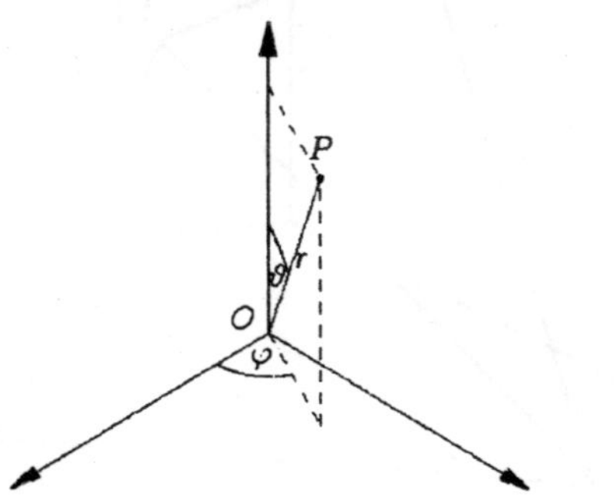

We have

$$x = r\cos(\varphi)\sin(\vartheta),\ y = r\sin(\varphi)\sin(\vartheta),\ z = r\cos(\vartheta)$$

and,

$$r = \sqrt{x^2 + y^2 + z^2},$$

$$\varphi = \arctan(y/x),$$

$$\vartheta = \arctan\sqrt{x^2 + y^2}/z^2),$$

but you have got to select the appropriate branch of depending on the signs of and: π

$x \geq 0 \wedge y \geq 0$	$\varphi \in [0, \pi/2)$
$x \leq 0$	$\varphi \in [\pi/2, 3\pi/2)$
$x \geq 0 \wedge y < 0$	$\varphi \in [3\pi/2, 2\pi)$
$z \leq 0$	$\vartheta \in [\pi/2, \pi)$

In Cartesian coordinates the points,

$P_1 = (1, 1, 1)$,
$P_2 = (1, -1, -1)$,
$P_3 = (-1, 1, -1)$,
$P_4 = (-1, -1, 1)$,

form the vertices of a regular tetrahedron with edge length $2\sqrt{2}$ and a barycentre which lies in the origin O.

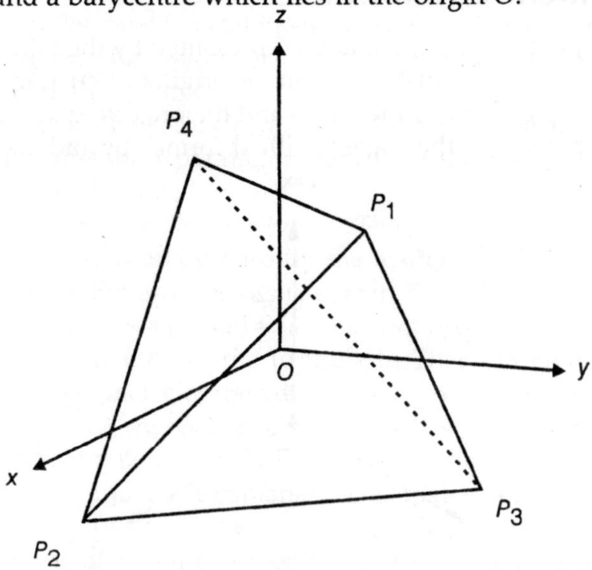

Represented by spherical coordinates (r, ϑ, φ) the vertices are given by,

$$P_1' = (\sqrt{3}, \psi, \pi/4), \qquad P_1' = (\sqrt{3}, \pi - \psi, \pi/4),$$

$$P_3' = (\sqrt{3}, \pi, -\psi, 3\pi/4), \qquad P_4' = (\sqrt{3}, \psi, -3\pi/4),$$

with $\psi = \arccos(1/\sqrt{3})$ respectively $\psi = \arctan\sqrt{2}$.

VECTORS IN EUCLIDEAN SPACE

In single-variable calculus, the functions that one encounters are functions of a variable (usually x or t) that varies over some subset of the real number line (which we denote by $\mathbb{R}$.). For such a function, say, $y = f(x)$, the graph of the function f consists of the points $(x, y) = (x, f(x))$. These points lie in the Euclidean plane, which, in the Cartesian or rectangular coordinate system, consists of all ordered pairs of real numbers (a, b).

We use the word "Euclidean" to denote a system in which all the usual rules of Euclidean geometry hold. We denote the Euclidean plane by $\mathbb{R}^2$; the "2" represents the number of *dimensions* of the plane.

The Euclidean plane has two perpendicular coordinate axes: the x-axis and the y-axis.

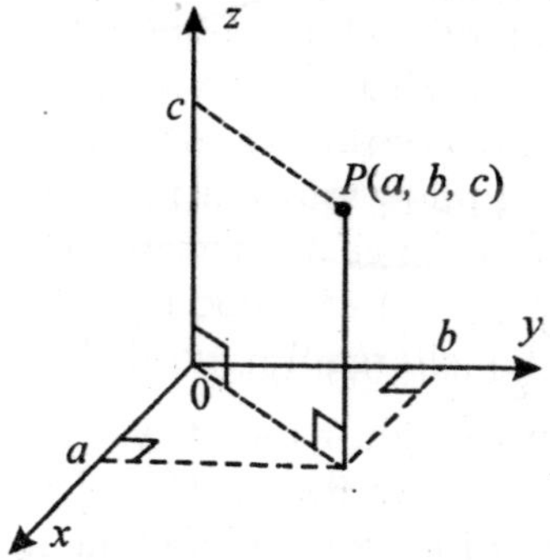

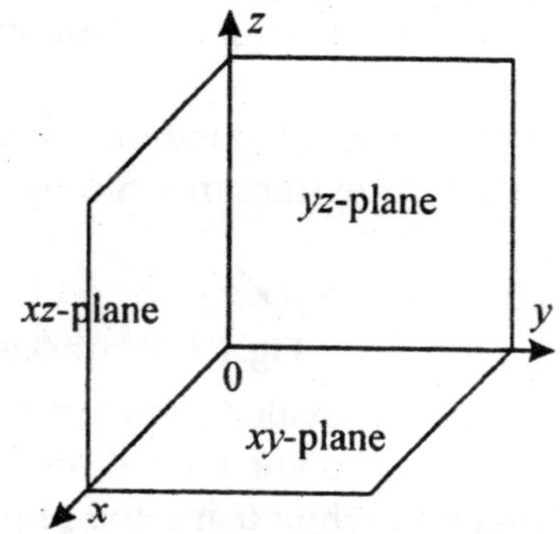

In vector (or multivariable) calculus, we will deal with functions of two or three variables (usually x, y or x, y, z, respectively). The graph of a function of two variables, say, $z = f(x, y)$, lies in Euclidean space, which in the Cartesian coordinate systemconsists of all ordered triples of real numbers (a, b, c). Since Euclidean space is 3-dimensional, we denote it by $\mathbb{R}^3$.

The graph of f consists of the points $(x, y, z) = (x, y, f(x, y))$. The 3-dimensional coordinate system of Euclidean space can be represented on a flat surface, such as this page or a blackboard, only by giving the illusion of three dimensions. Euclidean space has three mutually perpendicular coordinate axes (x, y and z), and three mutually perpendicular coordinate planes: the xy-plane, yz-plane and xz-plane.

The coordinate system shown in Figure is known as a right-handed coordinate system, because it is possible, using the right hand, to point the index finger in the positive direction of the x-axis, the middle finger in the positive direction of the y-axis, and the thumb in the positive direction of the z-axis.

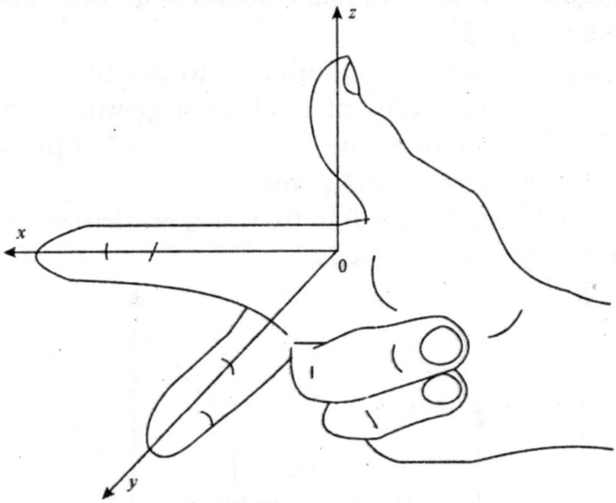

Fig. Right-handed Coordinate System

An equivalent way of defining a right-handed system is if you can point your thumb upwards in the positive z-axis direction while using the remaining four fingers to rotate the x-axis towards the y-axis.

Doing the same thing with the left hand is what defines a left-handed coordinate system. Notice that switching the x- and y-axes in a right-handed system results in a left-handed system, and that rotating either type of system does not change its "handedness". Throughout the book we will use a right-handed system.

For functions of three variables, the graphs exist in 4-dimensional space (*i.e.* R^4), which we can not see in our 3-dimensional space, let alone simulate in 2-dimensional space. So we can only think of 4-dimensional space abstractly. So far, we have discussed the *position* of an object in 2-dimensional or 3-dimensional space. But what about something such as the velocity of the object, or its acceleration? Or the gravitational force acting on the object? These phenomena all seem to involve motion and *direction* in some way. This is where the idea of a *vector* comes in. You have already dealt with velocity and acceleration in single-variable calculus.

For example, for motion along a straight line, if $y = f(t)$ gives the displacement of an object after time t, then $dy/dt = f'(t)$ is the velocity of the object at time t. The derivative $f'(t)$ is just a number, which is positive if the object is moving in an agreedupon "positive" direction, and negative if it moves in the opposite of that direction.

So you can think of that number, which was called the velocity of the object, as having two components: a *magnitude,* indicated by a nonnegative number, preceded by a *direction,* indicated by a plus or minus symbol (representing motion in the positive direction or the negative direction, respectively), *i.e.* $f'(t) = \pm a$ for some number $a \geq 0$. Then a is the magnitude of the velocity (normally called the *speed* of the object), and the $\pm$ represents the direction of the velocity (though the + is usually omitted for the positive direction).

For motion along a straight line, *i.e.* in a 1-dimensional space, the velocities are also contained in that 1-dimensional space, since they are just numbers. For general motion along a curve in 2- or 3-dimensional space, however, velocity will need to be represented by a multidimensional object which should have both a magnitude and a direction.

A geometric object which has those features is an arrow, which in elementary geometry is called a "directed line segment". This is the motivation for how we will define a vector.

Definition: A (nonzero) vector is a directed line segment drawn from a point P (called its initial point) to a point Q (called its terminal point), with P and Q being distinct points.

The vector is denoted by $\overrightarrow{PQ}$ ℝ Its magnitude is the length of the line segment, denoted by $\left\|\overrightarrow{PQ}\right\|$, and its direction is the same as that of the directed line segment. The zero vector is just a point, and it is denoted by 0.

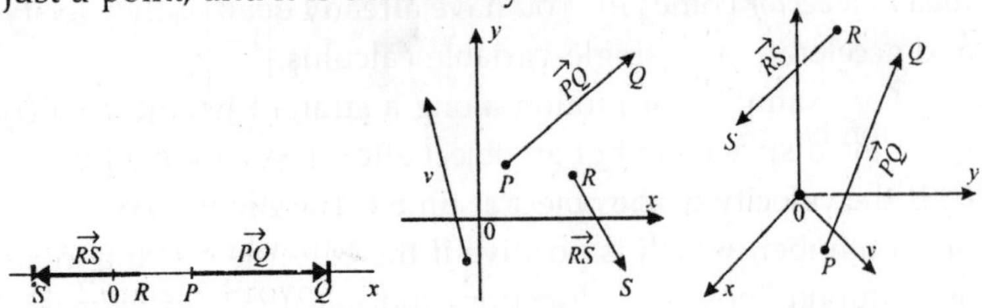

(a) One dimension (b) Two dimensions (c) Three dimensions

Fig. Vectors in Different Dimensions

To indicate the direction of a vector, we draw an arrow from its initial point to its terminal point. We will often denote a vector by a single bold-faced letter (*e. g. v*) and use the terms "magnitude" and "length" interchangeably. Note that our definition could apply to systems with any number of dimensions.

A few things need to be noted about the zero vector. Our motivation for what a vector is included the notions of magnitude and direction. What is the magnitude of the zero vector? We define it to be zero, *i.e.* || 0 || = 0. This agrees with the definition of the zero vector as just a point, which has zero length.

What about the direction of the zero vector? A single point really has no well-defined direction. Notice that we were careful to only define the direction of a *nonzero* vector, which is well-defined since the initial and terminal points are distinct. Not everyone agrees on the direction of the zero vector. Some contend that the zero vector has *arbitrary* direction (*i.e.* can take any direction), some say that it has *indeterminate* direction (*i.e.* the direction can not be determined), while others say that it has *no* direction. Our definition of the zero vector, however, does not require it to have a direction, and we will leave it at that.

Now that we know what a vector is, we need a way of determining when two vectors are equal. This leads us to the following definition.

Definition: Two nonzero vectors are equal if they have the same magnitude and the same direction. Any vector with zero magnitude is equal to the zero vector.

By this definition, vectors with the same magnitude and direction but with different initial points would be equal. For example, the vectors u, v and w all have the same magnitude $\sqrt{5}$ (by the Pythagorean Theorem). And we see that u and w are parallel, since they lie on lines having the same slope $\frac{1}{2}$, and they point in the same direction. So $u = w$, even though they have different initial points. We also see that v is parallel to u but points in the opposite direction. So u , v.

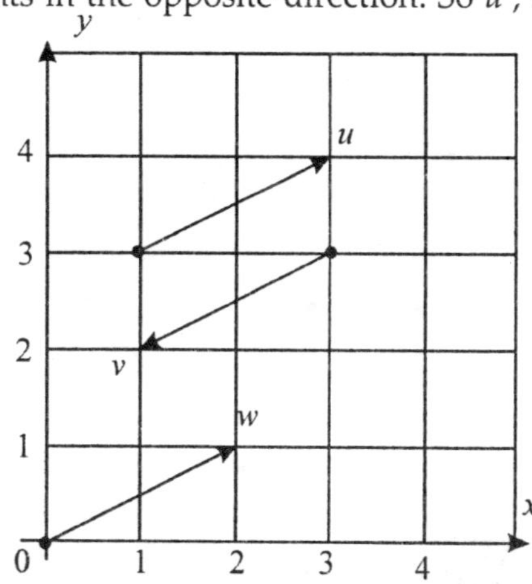

So we can see that there are an infinite number of vectors for a given magnitude and direction, those vectors all being equal and differing only by their initial and terminal points. Is there a single vector which we can choose to represent all those equal vectors?

Unless otherwise indicated, when speaking of "the vector" with a given magnitude and direction, we will mean the one whose initial point is at the origin of the coordinate system. Thinking of vectors as starting from the origin provides a way of dealing with vectors in a standard way, since every coordinate system has an origin. But there will be times when it is convenient to consider a different initial point for a vector (for example, when adding vectors, which we will do in the next section).

Another advantage of using the origin as the initial point is that it provides an easy correspondence between a vector and its terminal point.

Example: Let v be the vector in $\mathbb{R}^3$ whose initial point is at the origin and whose terminal point is (3, 4, 5). Though the *point* (3, 4, 5) and the vector v are different objects, it is convenient to write v = (3, 4, 5). When doing this, it is understood that the initial point of v is at the origin (0, 0, 0) and the terminal point is (3, 4, 5).

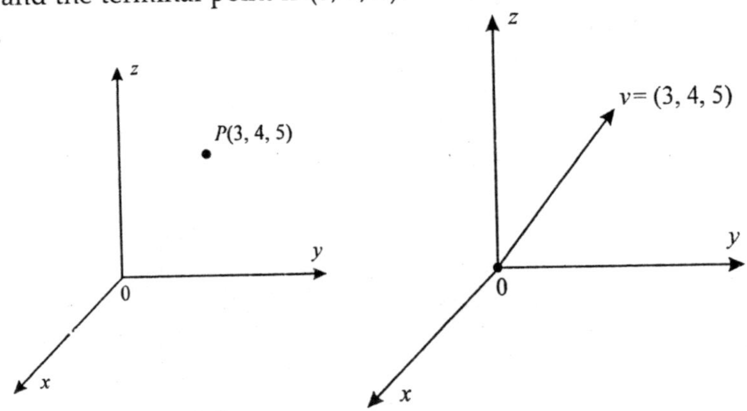

(a) The point (3, 4, 5)

(b) The vector (3, 4, 5)

Fig. Correspondence between points and vectors

Unless otherwise stated, when we refer to vectors as v = (a, b) in $\mathbb{R}^2$ or $v = (a, b, c)$ in $\mathbb{R}^3$, we mean vectors in Cartesian coordinates starting at the origin. Also, we will write the zero vector 0 in $\mathbb{R}^2$ and $\mathbb{R}^3$ as (0, 0) and (0, 0, 0), respectively.

The point-vector correspondence provides an easy way to check if two vectors are equal, without having to determine their magnitude and direction. Similar to seeing if two points are the same, you are now seeing if the terminal points of vectors starting at the origin are the same.

For each vector, find the (unique!) vector it equals whose initial point is the origin. Then compare the coordinates of the terminal points of these "new" vectors: if those coordinates are the same, then the original vectors are equal. To get the "new" vectors starting at the origin, you *translate* each vector to start at the origin by subtracting the coordinates of the original initial point from the original terminal point. The resulting point will be the terminal point of the "new" vector whose initial point is the origin. Do this for each original vector then compare.

Example: Consider the vectors $\overrightarrow{PQ}$ and $\overrightarrow{RS}$ in $\mathbb{R}_3$, where $P = (2, 1, 5)$, $Q = (3, 5, 7)$, $R = (1, -3, -2)$ and $S = (2, 1, 0)$. Does $\overrightarrow{PQ} = \overrightarrow{RS}$?

Solution: The vector $\overrightarrow{PQ}$ is equal to the vector v with initial point (0, 0, 0) and terminal point $Q-P = (3, 5, 7)-(2, 1, 5) = (3-2, 5-1, 7-5) = (1, 4, 2)$.

Similarly, $\overrightarrow{RS}$ is equal to the vector w with initial point (0, 0, 0) and terminal point $S-R = (2, 1, 0)-(1, -3, -2) = (2-1, 1-(-3), 0-(-2)) = (1, 4, 2)$.

So $\overrightarrow{PQ} = v = (1, 4, 2)$ and $\overrightarrow{RS} = w = (1, 4, 2)$.

$\therefore \quad \overrightarrow{PQ} = \overrightarrow{RS}$

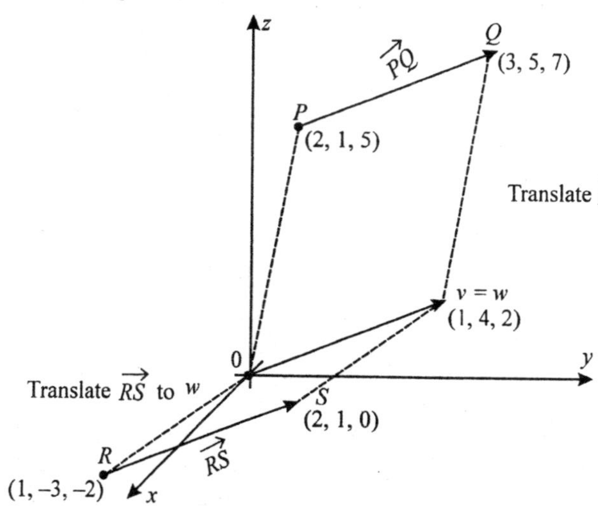

Recall the distance formula for points in the Euclidean plane:

For points $P = (x_1, y_1)$, $Q = (x_2, y_2)$ in $\mathbb{R}^2$, the distance d between P and Q is:

$$d = \sqrt{(x_2 - x_1)^2 + (y_2 - y_1)^2}$$

By this formula, we have the following result: For a vector $\overrightarrow{PQ}$ in $\mathbb{R}^2$ with initial point $P = (x_1, y_1)$ and terminal point $Q = (x_2, y_2)$, the magnitude of $\overrightarrow{PQ}$ is:

$$||\overrightarrow{PQ}|| = \sqrt{(x_2 - x_1)^2 + (y_2 - y_1)^2}$$

Finding the magnitude of a vector $v = (a, b)$ in $\mathbb{R}^2$ is a special case of formula with $P = (0, 0)$ and $Q = (a, b)$:

For a vector $v = (a, b)$ in $\mathbb{R}^2$, the magnitude of v is:

$$||v|| = \sqrt{a^2 + b^2}$$

To calculate the magnitude of vectors in $\mathbb{R}^3$, we need a distance formula for points in Euclidean space:

Theorem: The distance d between points $P = (x_1, y_1, z_1)$ and $Q = (x_2, y_2, z_2)$ in ¡^{3} is:

$$d = \sqrt{(x_2 - x_1)^2 + (y_2 - y_1)^2 (x_2 - z_1)^2}$$

The proof will use the following result:

Theorem: For a vector $v = (a, b, c)$ in R^3, the magnitude of v is:

$$||v|| = \sqrt{a^2 + b^2 + c^2}$$

Proof: There are four cases to consider:

Case 1: $a = b = c = 0$. Then $v = 0$, so $||\ v\ || = 0$

$$= \sqrt{0^2 + 0^2 + 0^2} = \sqrt{a^2 + b^2 + c^2}.$$

Case 2: Exactly two of a, b, c are 0. Without loss of generality, we assume that $a = b = 0$ and c ¹ 0 (the other two possibilities are handled in a similar manner). Then $v = (0, 0, c)$, which is a vector of length $|\ c\ |$ along the z-axis. So $||\ v\ || = |\ c\ |$

$$= \sqrt{c^2} = \sqrt{0^2 + 0^2 + c^2} = \sqrt{a^2 + b^2 + c^2}.$$

Case 3: Exactly one of a, b, c is 0. Without loss of generality, we assume that $a = 0$, b ¹ 0 and c ¹ 0 (the other two possibilities are handled in a similar manner). Then $v = (0, b, c)$, which is a vector in the yz-plane, so by the Pythagorean.

Theorem we have,

$$||\ v\ || = \sqrt{b^2 + c^2} = \sqrt{0^2 + b^2 + c^2} = \sqrt{a^2 + b^2 + c^2}$$

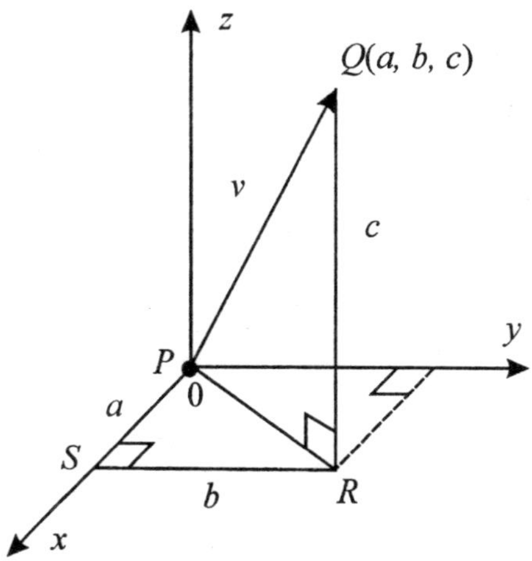

Case 4: None of a, b, c are 0. Without loss of generality, we can assume that a, b, c are all positive (the other seven possibilities are handled in a similar manner). Consider the points $P = (0, 0, 0)$, $Q = (a, b, c)$, $R = (a, b, 0)$, and $S = (a, 0, 0)$. Applying the Pythagorean Theorem to the right triangle DPS R gives $| PR |^2 = a^2 + b^2$. A second application of the Pythagorean Theorem, this time to the right triangle DPQR, gives,

$$|| v || = | PQ | = \sqrt{|PR|^2 + |QR|^2} = \sqrt{a^2 + b^2 + c^2}.$$

This proves the theorem.

Example: Calculate the following:

1. The magnitude of the vector $\overrightarrow{PQ}$ in ¡^{2} with $P = (-1, 2)$ and $Q = (5, 5)$.

   Solution: By formula,

   $$||\overrightarrow{PQ}|| = \sqrt{(5-(-1))^2 + (5-2)^2}$$

   $$= \sqrt{36+9}$$

   $$= \sqrt{45} = 3\sqrt{5}.$$

2. The magnitude of the vector $v = (8, 3)$ in R^2.

   Solution: By formula, $|| v || = \sqrt{8^2 + 3^2} = \sqrt{73}$.

3. The distance between the points $P = (2, -1, 4)$ and $Q = (4, 2, -3)$ in ¡^{2}.

Solution: By formula, the distance

$$d = \sqrt{(4-2)^2 + (2-(-1))^2 + (-3-4)^2}$$

$$= \sqrt{4+9+49} = \sqrt{62}.$$

4. The magnitude of the vector $v = (5, 8, -2)$ in $\mathbb{R}^3$.
Solution: By formula,

$$\| v \| = \sqrt{5^2 + 8^2 + (-2)^2}$$

$$= \sqrt{25+64+4} = \sqrt{93}.$$

2

Vector Algebra

Now that we know what vectors are, we can start to perform some of the usual algebraic operations on them (*e.g.* addition, subtraction). Before doing that, we will introduce the notion of a *scalar*.

Definition: A scalar is a quantity that can be represented by a single number.

For our purposes, scalars will always be real numbers. Examples of scalar quantities are mass, electric charge, and speed (not velocity).4 We can now define *scalar multiplication* of a vector.

Definition: For a scalar k and a nonzero vector v, the scalar multiple of v by k, denoted by kv, is the vector whose magnitude is $|k| \|v\|$, points in the same direction as v if $k > 0$, points in the opposite direction as v if $k < 0$, and is the zero vector 0 if $k = 0$. For the zero vector 0, we define $k0 = 0$ for any scalar k.

Two vectors v and w are parallel (denoted by $v \parallel w$) if one is a scalar multiple of the other. You can think of scalar multiplication of a vector as stretching or shrinking the vector, and as flipping the vector in the opposite direction if the scalar is a negative number.

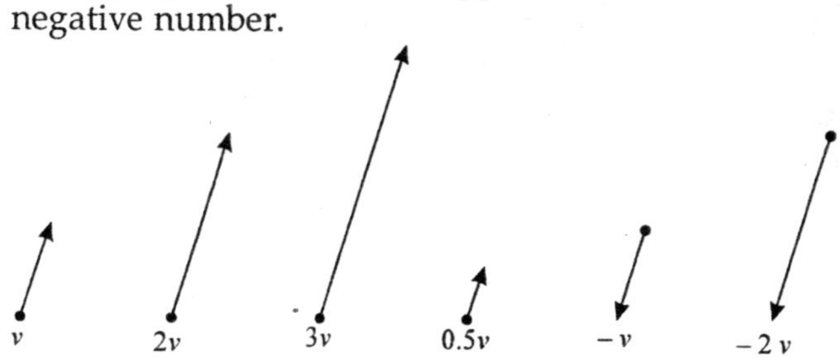

Recall that translating a nonzero vector means that the initial point of the vector is changed but the magnitude and direction are preserved. We are now ready to define the *sum* of two vectors.

Definition: The sum of vectors v and w, denoted by $v + w$, is obtained by translating w so that its initial point is at the terminal point of v; the initial point of $v + w$ is the initial point of v, and its terminal point is the new terminal point of w.

Intuitively, adding w to v means tacking on w to the end of v.

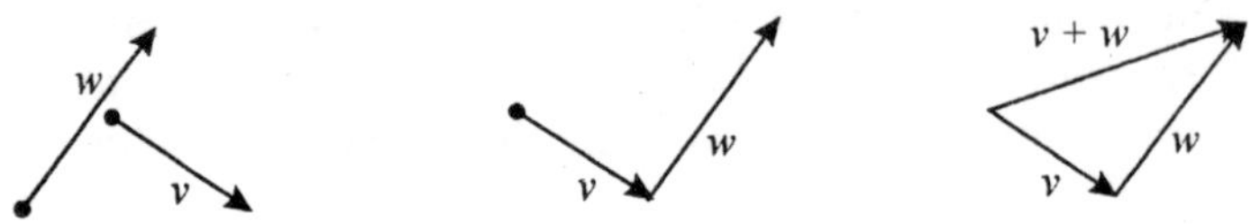

(a) Vectors v and w (b) Translate w to the end of v (c) The sum $v + w$

Fig. Adding Vectors v and w

Notice that our definition is valid for the zero vector (which is just a point, and hence can be translated), and so we see that $v + 0 = v = 0 + v$ for any vector v. In particular, $0 + 0 = 0$. Also, it is easy to see that $v + (-v) = 0$, as we would expect. In general, since the scalar multiple $-v = -1\ v$ is a well-defined vector, we can define vector subtraction as follows: $v - w = v + (-w)$.

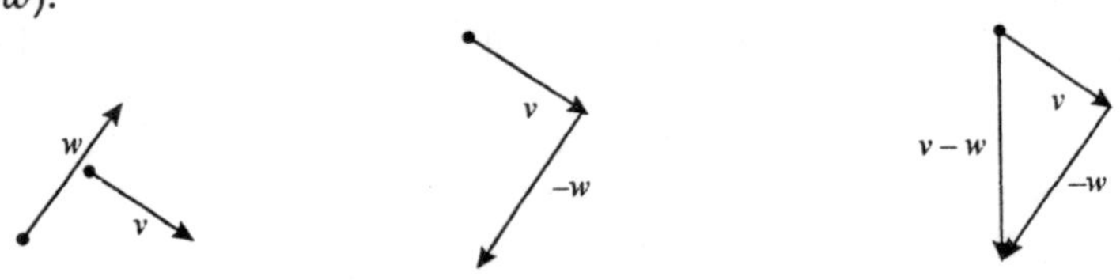

(a) Vectors v and w (b) Translate $-w$ to the end of v (c) The difference $v - w$

Fig. Subtracting vectors v and w

Figure shows the use of "geometric proofs" of various laws of vector algebra, that is, it uses laws from elementary geometry to prove statements about vectors. For example, (a) shows that $v + w = w + v$ for any vectors v, w. And (c) shows how you can think of $v - w$ as the vector that is tacked on to the end of w to add up to v.

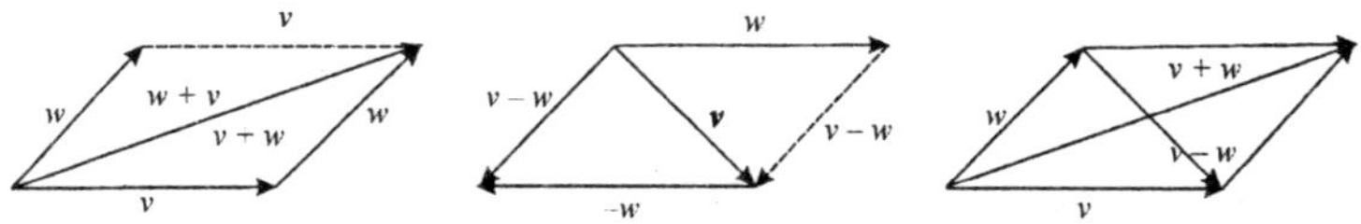

(a) Add vectors (b) Subtract vectors (c) Combined add/subtract

Fig. Geometric" Vector Algebra

Notice that we have temporarily abandoned the practice of starting vectors at the origin. In fact, we have not even mentioned coordinates in this section so far. Since we will deal mostly with Cartesian coordinates in this book, the following two theorems are useful for performing vector algebra on vectors in $\mathbb{R}^2$ and $\mathbb{R}^3$ starting at the origin.

Theorem: Let,

$$v = (v_1, v_2), w = (w_1, w_2)$$

be vectors in $\mathbb{R}^2$, and let k be a scalar.

Then,

(a) $kv = (kv_1, kv_2)$

(b) $v + w = (v_1 + w_1, v_2 + w_2)$

Proof: (a) Without loss of generality, we assume that $v_1, v_2 > 0$ (the other possibilities are handled in a similar manner). If, $k = 0$ then,

$$kv = 0v = 0 = (0, 0) = (0v_1, 0v_2) = (kv_1, kv_2),$$

which is what we needed to show.

If $k \neq 0$, then (kv_1, kv_2) lies on a line with slope,

$$\frac{kv_2}{kv_1} = \frac{v_2}{v_1},$$

which is the same as the slope of the line on which v (and hence kv) lies, and (kv_1, kv_2) points in the same direction on that line as kv.

Also, by formula the magnitude of (kv_1, kv_2) is $p(kv_1)$,

$$\sqrt{(kv_1)^2 + (kv_2)^2} = \sqrt{k^2 v_1^2 + k^2 v_2^2} = \sqrt{k^2(v_1^2 + v_2^2)} = | k |$$

$$\sqrt{v_1^2 + v_2^2} = | k | \; || v ||.$$

So kv and (kv_1, kv_2) have the same magnitude and direction. This proves (a).

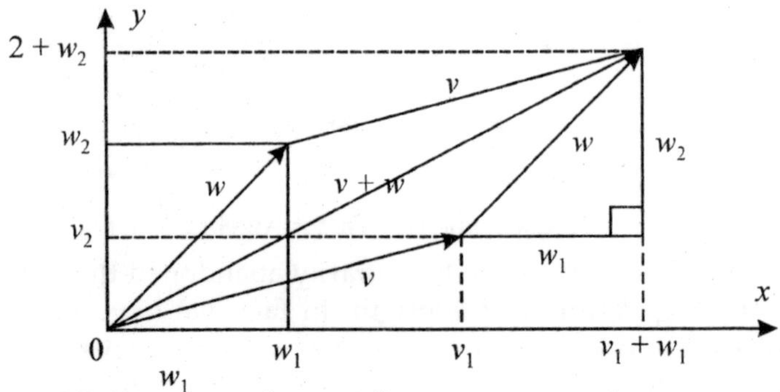

- Without loss of generality, we assume that v_1, v_2, w_1, $w_2 > 0$ (the other possibilities are handled in a similar manner). From Figure, we see that when translating w to start at the end of v, the new terminal point of w is $(v_1 + w_1, v_2 + w_2)$, so by the definition of $v + w$ this must be the terminal point of $v + w$. This proves (b).

Theorem: Let $v = (v_1, v_2, v_3)$, $w = (w_1, w_2, w_3)$ be vectors in $\mathbb{R}^3$, let k be a scalar. Then

(a) $kv = (kv_1, kv_2, kv_3)$

(b) $v + w = (v_1 + w_1, v_2 + w_2, v_3 + w_3)$

The following theorem summarizes the basic laws of vector algebra.

Theorem: For any vectors u, v, w, and scalars k, l, we have

(a) $v + w = w + v$ Commutative Law

(b) $u + (v + w) = (u + v) + w$ Associative Law

(c) $v + 0 = v = 0 + v$ Additive Identity

(d) $v + (-v) = 0$ Additive Inverse

(e) $k(lv) = (kl)v$ Associative Law

(f) $(v + w) = kv + kw$ Distributive Law

(g) $(k + l)v = kv + lv$ Distributive Law

Proof: (a) We already presented a geometric proof of this in Figure.

- To illustrate the difference between analytic proofs and geometric proofs in vector algebra, we will present both types here. For the analytic proof, we will use vectors in ¡^{3} (the proof for ¡^{2} is similar). Let,

  $u = (u_1, u_2, u_3)$,

  $v = (v_1, v_2, v_3)$, w

$= (w_1, w_2, w_3)$ be vectors in ¡^{3}.

Then,

$u + (v + w) = (u_1, u_2, u_3) + ((v_1, v_2, v_3) + (w_1, w_2, w_3))$

$= (u_1, u_2, u_3) + (v_1 + w_1, v_2 + w_2, v_3 + w_3)$ by Theorem

$= (u_1 + (v_1 + w_1), u_2 + (v_2 + w_2), u_3 + (v_3 + w_3))$ by Theorem

$= ((u_1 + v_1) + w_1, (u_2 + v_2) + w_2, (u_3 + v_3) + w_3)$ by properties of real numbers

$= (u_1 + v_1, u_2 + v_2, u_3 + v_3) + (w_1, w_2, w_3)$ by Theorem

$= (u + v) + w$

This completes the analytic proof of (b). Figure provides the geometric proof.

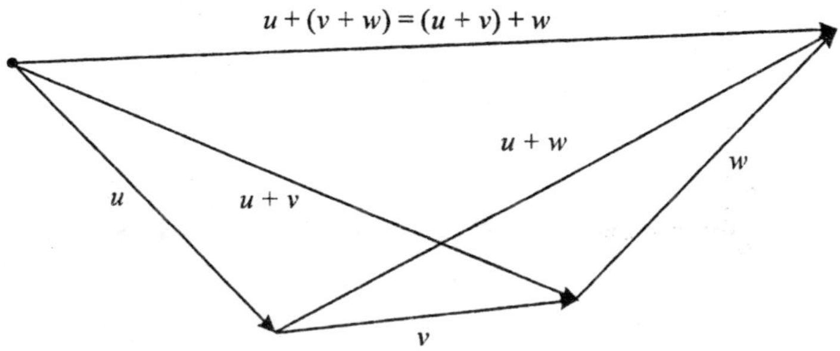

Fig. Associative Law for vector addition

- We already discussed.
- We already discussed.
- We will prove this for a vector $v = (v_1, v_2, v_3)$ in ¡^{3} (the proof for ¡^{2} is similar):

$k(lv) = k(lv_1, lv_2, lv_3)$ by Theorem

$= (klv_1, klv_2, klv_3)$ by Theorem

$= (kl)(v_1, v_2, v_3)$ by Theorem

$= (kl)v$

(f) and *(g)*: Left as exercises for the reader. A unit vector is a vector with magnitude 1. Notice that for any nonzero vector v, the vector,

$$\frac{v}{||v||}$$

is a unit vector which points in the same direction as

v, since,

$$\frac{1}{||v||} > 0 \text{ and } \left\| \frac{v}{||v||} \right\| = \frac{||v||}{||v||} = 1.$$

Dividing a nonzero vector v by $||\ v\ ||$ is often called *normalizing* v.

There are specific unit vectors which we will often use, called the basis vectors: $i = (1, 0, 0)$, $j = (0, 1, 0)$, and $k = (0, 0, 1)$ in $\mathbb{R}^3$; $i = (1, 0)$ and $j = (0, 1)$ in $\mathbb{R}^2$. These are useful for several reasons: they are mutually perpendicular, since they lie on distinct coordinate axes; they are all unit vectors: $||\ i\ || = ||\ j\ || = ||\ k\ || = 1$; every vector can be written as a unique scalar combination of the basis vectors: $v = (a, b) = a\,i + b\,j$ in $\mathbb{R}^2$, $v = (a, b, c) = a\,i + b\,j + ck$ in $\mathbb{R}^3$.

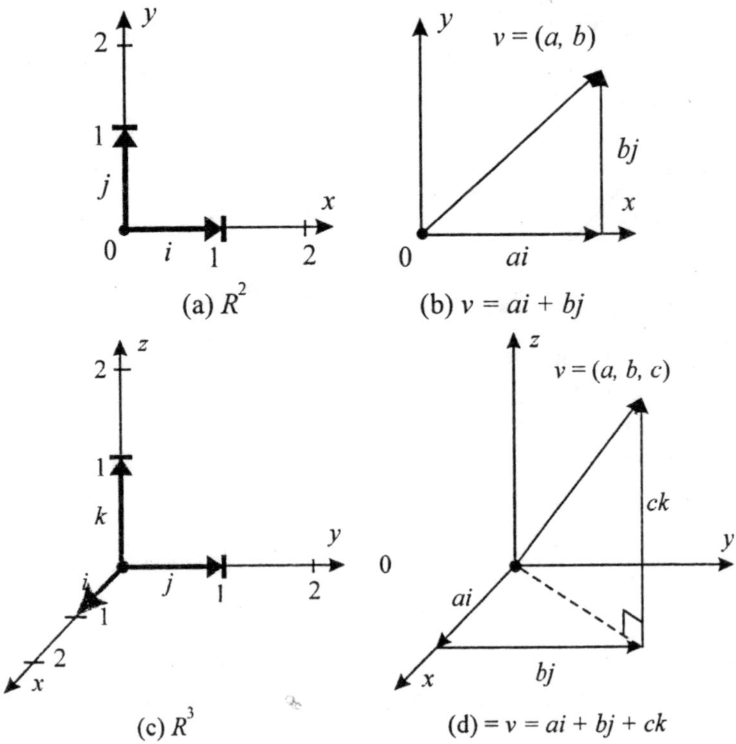

Fig. Basis Vectors in Different Dimensions

When a vector $v = (a, b, c)$ is written as $v = a\,i + b\,j + ck$, we say that v is in *component form*, and that a, b, and c are the *i, j, and k components*, respectively, of v.

We have:

$v = v_1\, i + v_2\, j + v_3\, k$, k a scalar Þ

$kv = kv_1\, i + kv_2\, j + kv_3\, k$

$v = v_1\, i + v_2\, j + v_3\, k,\ w$

$= w_1\, i + w_2\, j + w_3\, k$ Þ $v + w$

$= (v_1 + w_1)i + (v2 + w_2)j + (v_3 + w_3)k$

$v = v_1\, i + v_2\, j + v_3\, k$ Þ $||\, v\, ||$

$= \sqrt{v_1^2 + v_2^2 + v_3^2}$

Example: Let $v = (2, 1, -1)$ and $w = (3, -4, 2)$ in ¡^{3}.

(a) Find $v–w$.
Solution: $v–w = (2–3, 1–(–4)–1–2) = (–1, 5, –3)$

(b) Find $3v + 2w$.
Solution: $3v + 2w = (6, 3, –3) + (6, –8, 4) = (12, –5, 1)$

(c) Write v and w in component form.
Solution: $v = 2\, i + j–k$, $w = 3\, i–4\, j + 2k$

(d) Find the vector u such that $u + v = w$.
Solution: By Theorem,
$u = w–v = –(v–w) = –(–1, 5, –3) = (1, –5, 3)$.3

(e) Find the vector u such that $u + v + w = 0$.
Solution: By Theorem, $u = –w–v = –(3, –4, 2)–(2, 1, –1) = (–5, 3, –1)$.

(f) Find the vector u such that $2u + i–2\, j = k$.

Solution: $2u = –i + 2\, j + k \Rightarrow u = -\frac{1}{2}i + j + \frac{1}{2}k$

(g) Find the unit vector $\frac{v}{||v||}$.

Solution:

$$\frac{v}{||v||} = \frac{1}{\sqrt{2^2+1^2+(-1)^2}}(2,1,-1) = \left(\frac{2}{\sqrt{6}}, \frac{1}{\sqrt{6}}, \frac{-1}{\sqrt{6}}\right)$$

We can now easily prove Theorem from the previous section. The distance d between two points $P = (x_1, y_1, z_1)$ and $Q = (x_2, y_2, z_2)$ in $\mathbb{R}^3$ is the same as the length of the vector $w–v$, where the vectors v and w are defined as $v = (x_1, y_1, z_1)$ and $w = (x_2, y_2, z_2)$. So since, $w–v = (x_2–x_1, y_2–y_1, z_2–z_1)$, then,

$d = ||\, w–v\, ||$

$$= \sqrt{(x_2 - x_1)^2 + (y_2 - y_1)^2 + (z_2 - z_1)^2} \text{ by Theorem.}$$

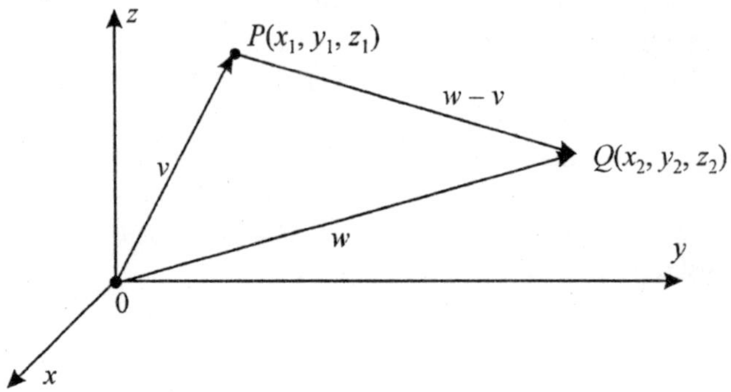

Fig. Proof of Theorem : $d = ||w-v||$

Dot Product

You may have noticed that while we did define multiplication of a vector by a scalar in the previous section on vector algebra, we did not define multiplication of a vector by a vector. We will now see one type of multiplication of vectors, called the *dot product.*

Definition: Let $v = (v_1, v_2, v_3)$ and $w = (w_1, w_2, w_3)$ be vectors in $\mathbb{R}^3$. The dot product of v and w, denoted by $v \cdot w$, is given by:

$$v \cdot w = v_1w_1 + v_2w_2 + v_3w_3$$

Similarly, for vectors $v = (v_1, v_2)$ and $w = (w_1, w_2)$ in $\mathbb{R}^2$, the dot product is:

$$v \cdot w = v_1w_1 + v_2w_2$$

Notice that the dot product of two vectors is a scalar, not a vector. So the associative law that holds for multiplication of numbers and for addition of vectors, does *not* hold for the dot product of vectors. Why? Because for vectors u, v, w, the dot product $u \cdot v$ is a scalar, and so $(u \cdot v) \cdot w$ is not defined since the left side of that dot product (the part in parentheses) is a scalar and not a vector. For vectors $v = v_1 i + v_2 j + v_3 k$ and $w = w_1 i + w_2 j + w_3 k$ in component form, the dot product is still $v \cdot w = v_1w_1 + v_2w_2 + v_3w_3$.

Also notice that we defined the dot product in an analytic way, *i.e.* by referencing vector coordinates.

There is a geometric way of defining the dot product, which we will now develop as a consequence of the analytic definition.

Definition: The angle between two nonzero vectors with the same initial point is the smallest angle between them. We do not define the angle between the zero vector and any other vector. Any two nonzero vectors with the same initial point have two angles between them: θ and 360°–θ. We will always choose the smallest nonnegative angle θ between them, so that $0° \leq \theta \leq 180°$.

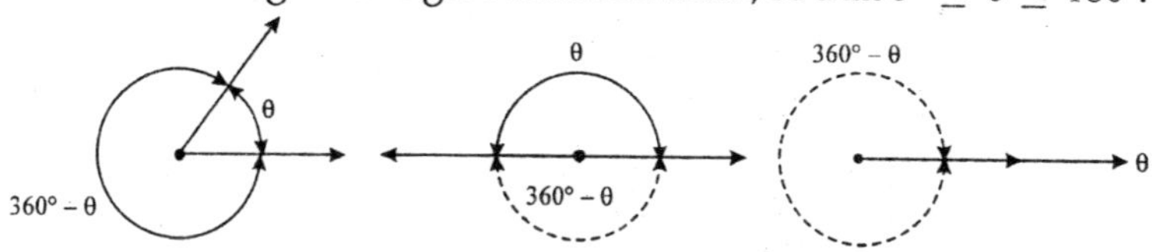

(a) 0° < θ < 180° (b) θ = 180° (c) θ = 0°

Fig. Angle between Vectors

We can now take a more geometric view of the dot product by establishing a relationship between the dot product of two vectors and the angle between them.

Theorem: Let *v*, *w* be nonzero vectors, and let θ be the angle between them. Then,

$$\cos\theta = \frac{v.w}{||v||\,||w||}$$

Proof: We will prove the theorem for vectors in ¡^{3} (the proof for ¡^{2} is similar). Let $v = (v_1, v_2, v_3)$ and $w = (w_1, w_2, w_3)$. By the Law of Cosines, we have,

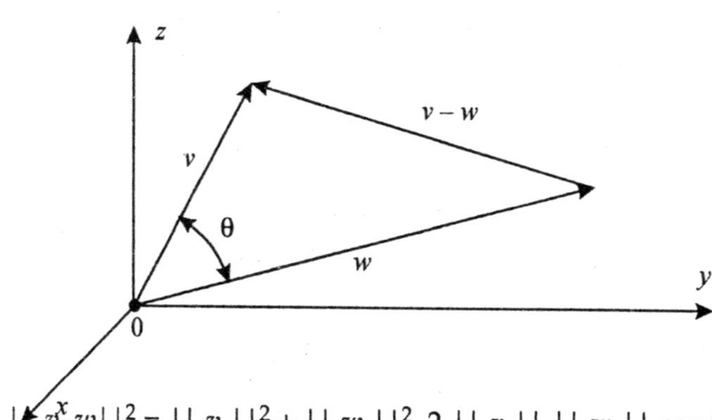

$$||v – w||^2 = ||v||^2 + ||w||^2 – 2\,||v||\,||w|| \cos\theta$$

(note that equation holds even for the "degenerate" cases θ = 0° and 180°). Since $v-w = (v_1-w_1, v_2-w_2, v_3-w_3)$, expanding $\| v-w \|^2$ in equation gives,

$$\| v \|^2 + \| w \|^2 - 2 \| v \| \| w \| \cos\theta$$
$$= (v_1-w_1)^2 + (v_2-w_2)^2 + (v_3-w_3)^2$$
$$= \left(v_1^2 - v_2^2 + v_3^2\right) + \left(w_1^2 + w_2^2 + w_3^2\right) -$$
$$2(v_1w_1 + v_2w_2 + v_2w_2 + v_3w_3)$$
$$= \|v\|^2 \|w\|^2 - 2(v.w), \text{ so}$$
$$-2\|v\|\|w\|\cos\theta = -2(v.w), \text{ so since } v\ {}^1\ 0 \text{ and } w\ {}^1\ 0 \text{ then,}$$

$$\cos\theta = \frac{v.w}{\|v\|\|w\|}, \text{ so since } v\ {}^1\ 0 \text{ and } w\ {}^1\ 0 \text{ then}$$

$$\cos\theta = \frac{v.w}{\|v\|\|w\|}, \text{ since } \| v \| > 0 \text{ and } \| w \| > 0.$$

Example: Find the angle θ between the vectors $v = (2, 1, -1)$ and $w = (3, -4, 1)$.

Solution: Since $v \,.\, w = (2)(3) + (1)(-4) + (-1)(1) = 1$, $\| v \| = \sqrt{6}$, and $\| w \| = \sqrt{26}$, then,

$$\cos\theta = \frac{v.w}{\|v\|\|w\|} = \frac{1}{\sqrt{6}\sqrt{26}} = \frac{1}{2\sqrt{39}} \approx 0.08 \Rightarrow \theta = 85.41°$$

Two nonzero vectors are perpendicular if the angle between them is 90°. Since cos 90° = 0, we have the following important corollary to Theorem:

Corollary: Two nonzero vectors v and w are perpendicular if and only if $v \,.\, w = 0$. We will write $v \perp w$ to indicate that v and w are perpendicular. Since $\cos\theta > 0$ for $0° \leq \theta < 90°$ and $\cos\theta < 0$ for $90° < \theta \leq 180°$, we also have:

Corollary: If θ is the angle between nonzero vectors v and w, then,

$$v.w \text{ is } \begin{cases} > 0 & \text{for } 0° \leq \theta < 90° \\ 0 & \text{for } \theta = 90° \\ < 0 & \text{for } 90° < \theta \leq 180° \end{cases}$$

By Corollary, the dot product can be thought of as a way of telling if the angle between two vectors is acute, obtuse, or a right angle, depending on whether the dot product is positive, negative, or zero, respectively.

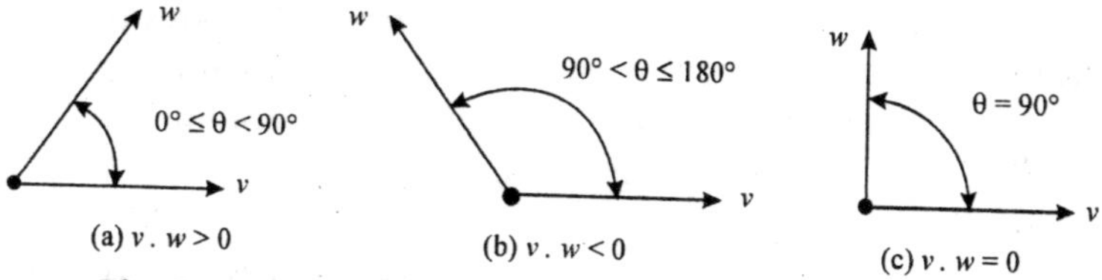

Fig. Sign of the dot Product and Angle between Vectors

Example: Are the vectors $v = (-1, 5, -2)$ and $w = (3, 1, 1)$ perpendicular?

Solution: Yes, $v \perp w$ since,

$$v \times w = (-1)(3) + (5)(1) + (-2)(1) = 0.$$

The following theorem summarizes the basic properties of the dot product.

Theorem: For any vectors u, v, w, and scalar k, we have

(a) $v \cdot w = w \cdot v$ Commutative Law

(b) $(kv) \cdot w = v \cdot (kw) = k(v \cdot w)$ Associative Law

(c) $v \cdot 0 = 0 = 0 \cdot v$

(d) $u \cdot (v + w) = u \cdot v + u \cdot w$ Distributive Law

(e) $(u + v) \cdot w = u \cdot w + v \cdot w$ Distributive Law

(f) $| v \cdot w | \leq || v || \, || w ||$ Cauchy-Schwarz Inequality[5]

Proof: The proofs of parts (a)–(e) are straightforward applications of the definition of the dot product, and are left to the reader as exercises. We will prove part (f). (f) If either $v = 0$ or $w = 0$, then $v \cdot w = 0$ by part (c), and so the inequality holds trivially. So assume that v and w are nonzero vectors. Then by Theorem,

$$v \cdot w = \cos\theta \, || v || \, || w ||, \text{ so}$$

$$| v \cdot w | = | \cos\theta | \, || v || \, || w ||, \text{ so}$$

$$| v \cdot w | \leq || v || \, || w || \text{ since } | \cos\theta | \leq 1.$$

Using Theorem we see that if $u \cdot v = 0$ and $u \cdot w = 0$, then $u \cdot (kv + lw) = k(u \cdot v) + l(u \cdot w) = k(0) + l(0) = 0$ for all scalars k, l. Thus, we have the following fact:

If $u \perp v$ and $u \perp w$, then $u \perp (kv + lw)$ for all scalars k, l. For vectors v and w, the collection of all scalar combinations $kv + lw$ is called the span of v and w. If nonzero vectors v and w are parallel, then their span is a line; if they are not parallel, then their span is a plane. So what we showed above is that a vector which is perpendicular to two other vectors is also perpendicular to their span.

The dot product can be used to derive properties of the magnitudes of vectors, the most important of which is the *Triangle Inequality*, as given in the following theorem:

Theorem: For any vectors v, w, we have,

(a) $|| v ||^2 = v \cdot v$

(b) $|| v + w || \leq || v || + || w ||$ Triangle Inequality

(c) $|| v - w || \geq || v || - || w ||$

Proof: (a) Left as an exercise for the reader.

(b) By part (a) and Theorem, we have $|| v + w ||^2 = (v + w) \cdot (v + w) = v \cdot v + v \cdot w + w \cdot v + w \cdot w$
$= || v ||^2 + 2(v \cdot w) + || w ||^2$, so since $a \leq | a |$ for any real number a, we have
$\leq || v ||^2 + 2 | v \cdot w | + || w ||^2$, so by Theorem we have $\leq || v ||^2 + 2 || v || \, || w || + || w ||^2 = (|| v || + || w ||)^2$ and so
$|| v + w || \leq || v || + || w ||$ after taking square roots of both sides, which proves (b).

(c) Since $v = w + (v - w)$, then $|| v || = || w + (v - w) || \leq || w || + || v - w ||$ by the Triangle Inequality, so subtracting $|| w ||$ from both sides gives $|| v || - || w || \leq || v - w ||$.

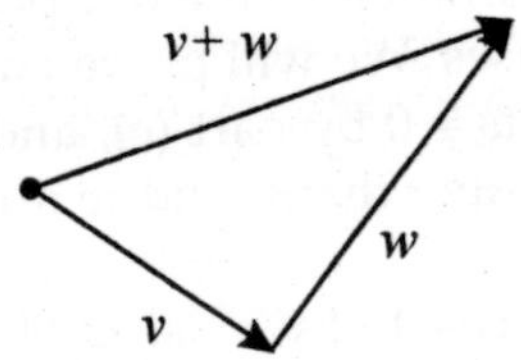

The Triangle Inequality gets its name from the fact that in any triangle, no one side is longer than the sum of the lengths of the other two sides. Another way of saying this is with the familiar statement "the shortest distance between two points is a straight line."

Cross Product

We have dcfined the dot product, which gave a way ofmultiplying two vectors. The resulting product, however, was a scalar, not a vector. In this section we will define a product of two vectors that does result in another vector. This product, called the *cross product,* is only defined for vectors in $\mathbb{R}^3$. The definition may appear strange and lacking motivation, but we will see the geometric basis for it shortly.

Definition: Let $v = (v_1, v_2, v_3)$ and $w = (w_1, w_2, w_3)$ be vectors in $\mathbb{R}^3$. The cross product of v and w, denoted by $v \,.\, w$, is the vector in $\mathbb{R}^3$ given by:

$$v \,.\, w = (v_2w_3 - v_3w_2,\ v_3w_1 - v_1w_3,\ v_1w_2 - v_2w_1)$$

Example: Find $i \times j$.

Solution: Since $i = (1, 0, 0)$ and $j = (0, 1, 0)$, then

$$i \times j = ((0)(0)-(0)(1),\ (0)(0)-(1)(0),\ (1)(1)-(0)(0))$$
$$= (0, 0, 1) = k$$

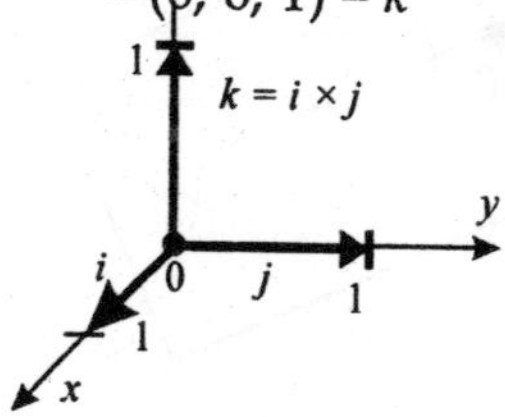

Similarly it can be shown that $j \,.\, k = i$ and $k \,.\, i = j$. In the above example, the cross product of the given vectors was perpendicular to both those vectors. It turns out that this will always be the case.

Theorem: If the cross product $v \,.\, w$ of two nonzero vectors v and w is also a nonzero vector, then it is perpendicular to both v and w.

Proof: We will show that $(v \,.\, w) \,.\, v = 0$: $(v \,.\, w)$.

$v = (v_2w_3 - v_3w_2,\ v_3w_1 - v_1w_3,\ v_1w_2 - v_2w_1) \,.\, (v_1, v_2, v_3)$

$= v_2w_3v_1 - v_3w_2v_1 + v_3w_1v_2 - v_1w_3v_2 + v_1w_2v_3 - v_2w_1v_3$

$= v_1v_2w_3 - v_1v_2w_3 + w_1v_2v_3 - w_1v_2v_3 + v_1w_2v3 - v_1w_2v_3$

$= 0$, after rearranging the terms.

$\therefore \quad v \times w \perp v$ by Corollary.

The proof that $v \times w \perp w$ is similar.

Corollary: If the cross product $v \times w$ of two nonzero vectors v and w is also a nonzero vector, then it is perpendicular to the span of v and w. The span of any two nonzero, nonparallel vectors v, w in $\mathbb{R}^3$ is a plane P, so the above corollary shows that $v \,.\, w$ is perpendicular to that plane.

As shown in Figure, there are two possible directions for $v \times w$, one the opposite of the other. It turns out that the direction of $v \times w$ is given by the *right-hand rule*, that is, the vectors v, w, $v \times w$ form a right-handed system. Recall that this means that you can point your thumb upwards in the direction of $v \times w$ while rotating v towards w with the remaining four fingers.

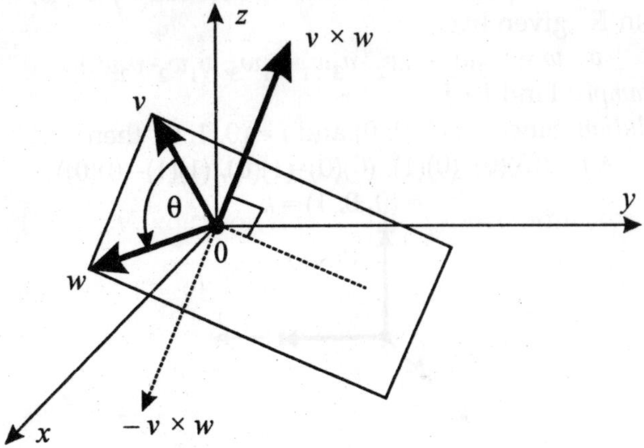

Fig. Direction of $v \times w$

We will now derive a formula for the magnitude of $v \times w$, for nonzero vectors v, w: $\| v \times w \|^2$

$$= (v_2w_3 - v_3w_2)^2 + (v_3w_1 - v_1w_3)^2 + (v_1w_2 - v_2w_1)^2$$

$$= v_2^2w_3^2 - 2v_2w_2v_3w_3 + v_3^2w_2^2 + v_3^2w_1^2$$

$$-2v_1w_1v_3w_3 + v_1^2w_3^2 + v_1^2w_2^2 - 2v_1w_1v_2w_2 + v_2^2w_1^2$$

$$= v_1^2\left(w_2^2 + w_3^2\right) + w_2^2\left(v_1^2 + w_3^2\right) + v_3^2\left(w_1^2 + w_2^2\right)$$

$$-2(v_1w_1v_2w_2 + v_1w_1v_3w_3 + v_2w_2v_3w_3)$$

and now adding and subtracting $v_1^2w_1^2, v_2^2w_2^2$, and $v_3^2w_3^2$ on the right side gives

$$= v_1^2\left(w_1^2 + w_2^2 + w_3^2\right) + v_2^2\left(w_1^2 + w_2^2 + w_3^2\right) + v_3^2\left(w_1^2 + w_2^2 + w_3^2\right) -$$

$$\left(v_1^2w_1^2 + v_2^2w_2^2 + v_3^2w_3^2 + 2(v_1w_1v_2w_2 + v_1w_1v_3w_3 + v_2w_2v_2w_3)\right)$$

$$-((v_1w_1)^2 + (v_2w_2)^2 + (v_3w_3)^2$$

$$+2(v_1w_1)(v_2w_2) + 2(v_1w_1)(v_3w_3) + 2(v_2w_2)(v_3w_3))$$

so using $(a + b + c)^2 = a^2 + b^2 + c^2 + 2ab + 2ac + 2bc$ for the subtracted term gives,

$$= \left(v_1^2 + v_2^2 + v_3^2\right)\left(w_1^2 + w_2^2 + w_3^2\right) - (v_1w_1 + v_2w_2 + v_3w_3)^2$$

$$= \| v \|^2 \| w \|^2 - (v \cdot w)^2$$

$$= \| v \|^2 \| w \|^2 \left(1 - \frac{(v.w)^2}{\|v\|^2 \|w\|^2}\right), \text{ since } \| v \| > 0$$

and $\| w \| > 0$, so by Theorem, $= \| v \|^2 \| w \|^2 (1 - \cos^2 \theta)$, where θ is the angle between v and w, so $\| v \cdot w \|^2 = \| v \|^2 \| w \|^2 \sin^2 \theta$, and since $0° \le \theta \le 180°$, then $\sin \theta \ge 0$, so we have: If θ is the angle between nonzero vectors v and w in $\mathbb{R}^3$, then,

$$\| v \cdot w \| = \| v \| \, \| w \| \sin \theta$$

It may seem strange to bother with the above formula, when the magnitude of the cross product can be calculated directly, like for any other vector. The formula is more useful for its applications in geometry, as in the following example.

Example: Let ΔPQR and $PQRS$ be a triangle and parallelogram, respectively.

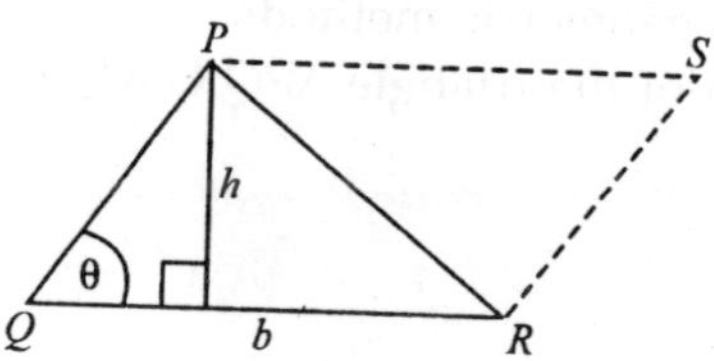

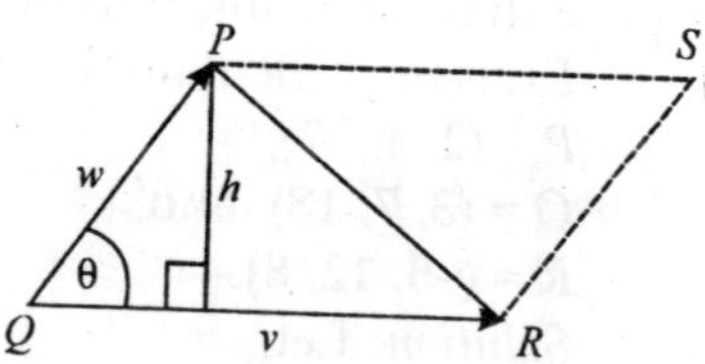

Think of the triangle as existing in $\mathbb{R}^3$, and identify the sides QR and QP with vectors v and w, respectively, in $\mathbb{R}^3$. Let θ be the angle between v and w. The area A_{PQR} of ΔPQR is $\frac{1}{2} bh$, where b is the base of the triangle and h is the height. So we see that,

$$b = \| v \| \text{ and } h = \| w \| \sin \theta$$

$$A_{PQR} = \frac{1}{2} \| v \| \, \| w \| \sin \theta$$

$$= \frac{1}{2} \| v \times w \|$$

So since the area A_{PQRS} of the parallelogram $PQRS$ is twice the area of the triangle ΔPQR, then,

$$A_{PQRS} = \| v \| \, \| w \| \sin \theta$$

By the discussion in Example, we have proved the following theorem:

Theorem: Area of triangles and parallelograms

(a) The area A of a triangle with adjacent sides v, w (as vectors in ¡^{3}) is:

$$A=\frac{1}{2}||v\times w||$$

(b) The area A of a parallelogram with adjacent sides v, w (as vectors in ¡^{3}) is:

$$A = || v \times w ||$$

It may seem at first glance that since the formulas derived in Example were for the adjacent sides QP and QR only, then the more general statements in Theorem that the formulas hold for *any* adjacent sides are not justified. We would get a different *formula* for the area if we had picked PQ and PR as the adjacent sides, but it can be shown that the different formulas would yield the same value, so the choice of adjacent sides indeed does not matter, and Theorem is valid. Theorem makes it simpler to calculate the area of a triangle in 3-dimensional space than by using traditional geometric methods.

Example: Calculate the area of the triangle ΔPQR, where,
$P = (2, 4, -7)$,
$Q = (3, 7, 18)$, and
$R = (-5, 12, 8)$.

Solution: Let,
$v = \overrightarrow{PQ}$ and
$w = \overrightarrow{PR}$, as in Figure.
Then,
$v = (3, 7, 18)-(2, 4, -7)$
$= (1, 3, 25)$ and
$w = (-5, 12, 8)-(2, 4, -7)$
$= (-7, 8, 15)$, so the area A of the triangle ΔPQR is,

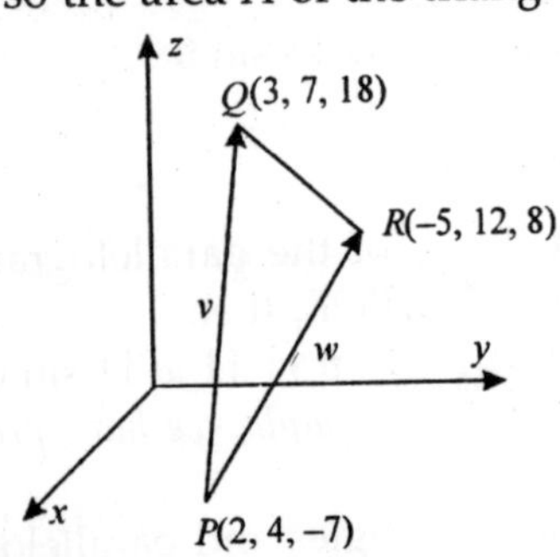

$$A = \frac{1}{2}||v \times w|| = \frac{1}{2}||(1,3,25) \times (-7 \times 8 \times 15)||$$
$$= \frac{1}{2}||((3)(15)-(25)(8),(28)(-7)-(1)(15),(1)(8)-(3)(-7)||$$
$$= \frac{1}{2}||(-155,-190,29)||$$
$$= \frac{1}{2}\sqrt{(-155)^2+(-190)^2+29^2} = \frac{1}{2}\sqrt{60966}$$
$$A \approx 123.46$$

Example: Calculate the area of the parallelogram $PQRS$, where $P = (1, 1)$, $Q = (2, 3)$, $R = (5, 4)$, and $S = (4, 2)$.

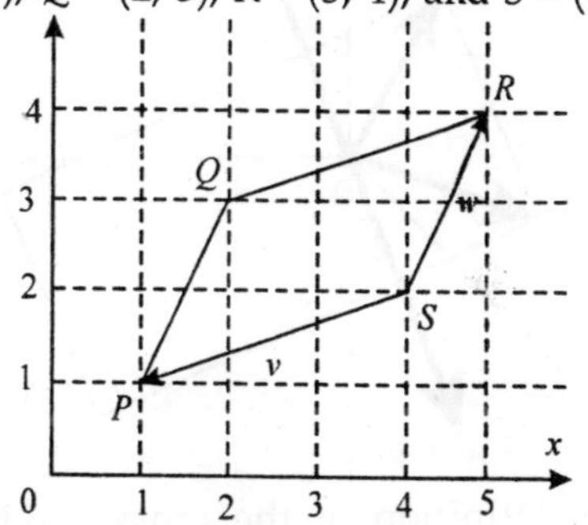

Solution: Let $v = \overrightarrow{SP}$ and $w = \overrightarrow{SR}$, as in Figure. Then $v = (1, 1)-(4, 2) = (-3, -1)$ and $w = (5, 4)-(4, 2) = (1, 2)$. But these are vectors in $\mathbb{R}^2$, and the cross product is only defined for vectors in $\mathbb{R}^3$. However, $\mathbb{R}^2$ can be thought of as the subset of $\mathbb{R}^3$ such that the z-coordinate is always 0. So we can write $v = (-3, -1, 0)$ and $w = (1, 2, 0)$. Then the area A of $PQRS$ is

$$A = || v \times w ||$$
$$= || (-3, -1, 0) \times (1, 2, 0)$$
$$= ((-1)(0)-(0)(2), (0)(1)-(-3)(0), (-3)(2)-(-1)(1))||$$
$$= || (0, 0, -5) ||$$
$$A = 5$$

The following theorem summarizes the basic properties of the cross product.

Theorem: For any vectors u, v, w in $\mathbb{R}^3$, and scalar k, we have,

(a) $v \,.\, w = -w \,.\, v$ Anticommutative Law

(b) $u \,.\, (v + w) = u \,.\, v + u \,.\, w$ Distributive Law

(c) $(u + v) \,.\, w = u \,.\, w + v \,.\, w$ Distributive Law

(d) $(kv) \,.\, w = v \,.\, (kw) = k(v \,.\, w)$ Associative Law

(e) $v \,.\, 0 = 0 = 0 \,.\, v$

(f) $v \,.\, v = 0$

(g) $v \,.\, w = 0$ if and only if $v \,||\, w$

Proof: The proofs of properties (b)–(f) are straightforward. We will prove parts (a) and (g) and leave the rest to the reader as exercises.

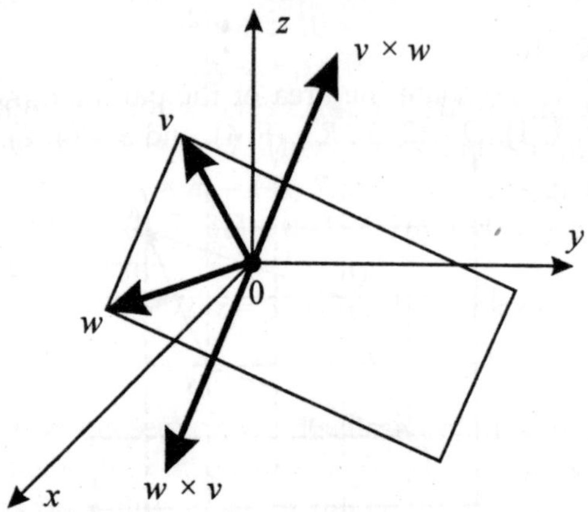

(a) By the definition of the cross product and scalar multiplication, *We have:*

$$\begin{aligned} v \times w &= (v_2w_3-v_3w_2,\ v_3w_1-v_1w_3,\ v_1w_2-v_2w_1) \\ &= -(v_3w_2-v_2w_3,\ v_1w_3-v_3w_1,\ v_2w_1-v_1w_2) \\ &= -(w_2v_3-w_3v_2,\ w_3v_1-w_1v_3,\ w_1v_2-w_2v_1) \\ &= -w \times v \end{aligned}$$

Note that this says that $v \times w$ and $w \times v$ have the same magnitude but opposite direction. (g) If either v or w is 0 then $v \times w = 0$ by part (e), and either $v = 0 = 0w$ or $w = 0 = 0v$, so v and w are scalar multiples, *i.e.* they are parallel.

If both v and w are nonzero, and θ is the angle between them, then by formula, $v \times w = 0$ if and only if $||\, v \,||\ ||\, w \,||\ \sin \theta = 0$, which is true if and only if $\sin \theta = 0$ (since $||\, v \,|| > 0$ and $||\, w \,|| > 0$). So since $0° \leq \theta \leq 180°$, then $\sin \theta = 0$ if and only if $\theta = 0°$ or $180°$. But the angle between v and w is $0°$ or $180°$ if and only if $v \,||\, w$.

Example: Adding to Example, we have

- $i \times j = k$
- $j \times k = i$
- $k \times i = j$
- $j \times i = -k$
- $k \times j = -i$
- $i \times k = -j$
- $i \times i = j \times j = k \times k = 0$

Example: Volume of a parallelepiped: Let the vectors *u, v, w* in $\mathbb{R}^3$ represent adjacent sides of a parallelepiped *P*, as in Figure. Show that the volume of *P* is the *scalar triple product u* . $(v \times w)$.

Solution: Recall that the volume of a parallelepiped is the area *A* of the base parallelogram times the height *h*. By Theorem, the area *A* of the base parallelogram is $\| v \times w \|$. And we can see that since $v \times w$ is perpendicular to the base parallelogram determined by *v* and *w*, then the height *h* is $\| u \| \cos\theta$, where θ is the angle between *u* and $v \times w$. By Theorem we know that,

$$\cos\theta = \frac{u.(v \times w)}{\|u\| \|v \times w\|}.$$

Hence,

$vol(P) = Ah$

$$= \| v \times w \| \frac{\|u\| u.(v \times w)}{\|u\| \|v \times w\|}$$

$= u \ . \ (v \times w)$

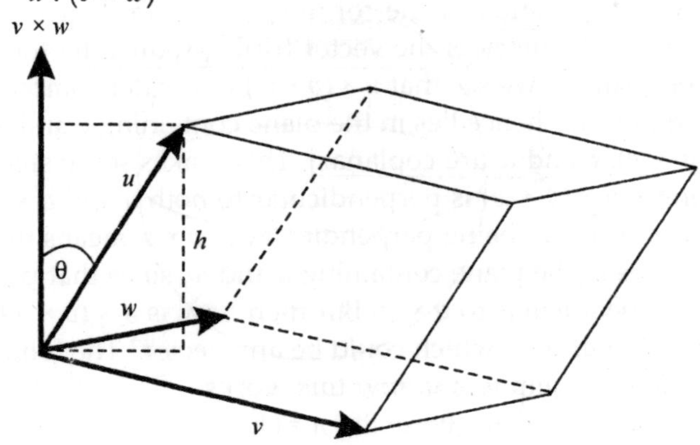

Fig. Parallelepiped *P*

In Example the height h of the parallelepiped is $\| u \| \cos \theta$, and not "$\| u \| \cos \theta$, because the vector u is on the same side of the base parallelogram's plane as the vector $v \times w$ (so that $\cos \theta > 0$).

Since the volume is the same no matter which base and height we use, then repeating the same steps using the base determined by u and v (since w is on the same side of that base's plane as $u \times v$), the volume is $w \, . \, (u \times v)$.

Repeating this with the base determined by w and u, we have the following result: For any vectors u, v, w in $\mathbb{R}^3$,

$$u \, . \, (v \times w) = w \, . \, (u \times v) = v \, . \, (w \times u)$$

(Note that the equalities hold trivially if any of the vectors are 0.) Since $v \times w = -w \times v$ for any vectors v, w in $\mathbb{R}^3$, then picking the wrong order for the three adjacent sides in the scalar triple product in formula will give you the negative of the volume of the parallelepiped. So taking the absolute value of the scalar triple product for *any* order of the three adjacent sides will *always* give the volume:

Theorem: If vectors u, v, w in $\mathbb{R}^3$ represent any three adjacent sides of a parallelepiped, then the volume of the parallelepiped is $| u \, . \, (v \times w) |$. Another type of triple product is the *vector triple product* $u \times (v \times w)$. The proof of the following theorem is left as an exercise for the reader:

Theorem: For any vectors u, v, w in $\mathbb{R}^3$,

$$u \times (v \times w) = (u \, . \, w)v - (u \, . \, v)w$$

An examination of the formula in Theorem gives some idea of the geometry of the vector triple product. By the right side of formula, we see that $u \times (v \times w)$ is a scalar combination of v and w, and hence lies in the plane containing v and w (*i.e.* $u \times (v \times w)$, v and w are coplanar). This makes sense since, by Theorem, $u \times (v \times w)$ is perpendicular to both u and $v \times w$.

In particular, being perpendicular to $v \times w$ means that $u \times (v \times w)$ lies in the plane containing v and w, since that plane is itself perpendicular to $v \times w$. But then how is $u \times (v \times w)$ also perpendicular to u, which could be any vector? The following example may help to see how this works.

Example. Find $u \times (v \times w)$ for $u = (1, 2, 4)$, $v = (2, 2, 0)$, $w = (1, 3, 0)$.

Solution: Since $u \cdot v = 6$ and $u \cdot w = 7$, then

$$u \times (v \times w) = (u \times w)v - (u \cdot v)w$$
$$= 7\,(2, 2, 0) - 6\,(1, 3, 0) = (14, 14, 0) - (6, 18, 0)$$
$$= (8, -4, 0)$$

Note that v and w lie in the xy-plane, and that $u \times (v \times w)$ also lies in that plane. Also, $u \times (v \times w)$ is perpendicular to both u and $v \times w = (0, 0, 4)$. For vectors $v = v_1\, i + v_2\, j + v_3\, k$ and $w = w_1\, i + w_2\, j + w_3\, k$ in component form, the cross product is written as: $v \times w = (v_2w_3 - v_3w_2)i + (v_3w_1 - v_1w_3)j + (v_1w_2 - v_2w_1)k$. It is often easier to use the component form for the cross product, because it can be represented as a *determinant*. We will not go too deeply into the theory of determinants7; we will just cover what is essential for our purposes.

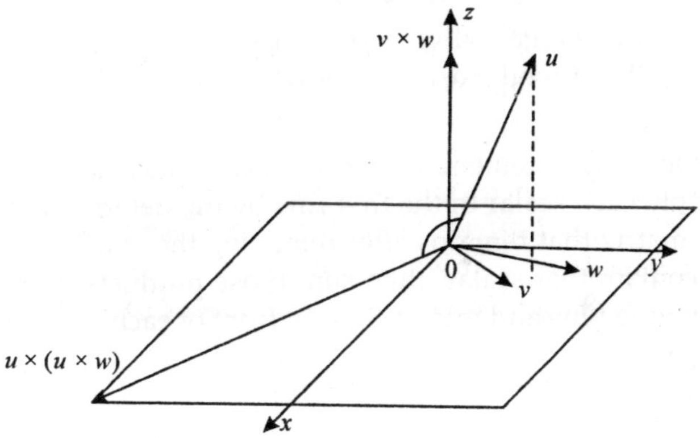

A 2 × 2 matrix is an array of two rows and two columns of scalars, written as,

$$\text{or} \begin{pmatrix} a & b \\ c & d \end{pmatrix}$$

where *a, b, c, d* are scalars. The determinant of such a matrix, written as

$$\begin{vmatrix} a & b \\ c & d \end{vmatrix} \text{ or det} \begin{bmatrix} a & b \\ c & d \end{bmatrix},$$

is the scalar defined by the following formula:

$$\begin{vmatrix} a & b \\ c & d \end{vmatrix} = ad - bc.$$

It may help to remember this formula as being the product of the scalars on the downward diagonal minus the product of the scalars on the upward diagonal.

Example:

$$\begin{vmatrix} 1 & 2 \\ 3 & 4 \end{vmatrix} = (1)(4) - (2)(3) = 4 - 6 = -2$$

A 3 × 3 matrix is an array of three rows and three columns of scalars, written as,

$$\begin{bmatrix} a_1 & a_2 & a_3 \\ b_1 & b_2 & b_3 \\ c_1 & c_2 & c_3 \end{bmatrix} \text{ or } \begin{pmatrix} a_1 & a_2 & a_3 \\ b_1 & b_2 & b_3 \\ c_1 & c_2 & c_3 \end{pmatrix}$$

and its determinant is given by the formula:

$$\begin{vmatrix} a_1 & a_2 & a_3 \\ b_1 & b_2 & b_3 \\ c_1 & c_2 & c_3 \end{vmatrix} = a_1 \begin{vmatrix} b_2 & b_3 \\ c_2 & c_3 \end{vmatrix} - a_2 \begin{vmatrix} b_1 & b_3 \\ c_1 & c_3 \end{vmatrix} + a_3 \begin{vmatrix} b_1 & b_2 \\ c_1 & c_2 \end{vmatrix}$$

One way to remember the above formula is the following: multiply each scalar in the first row by the determinant of the 2 × 2 matrix that remains after removing the row and column that contain that scalar, then sum those products up, putting alternating plus and minus signs in front of each (starting with a plus).

Example:

$$\begin{vmatrix} 1 & 0 & 2 \\ 4 & -1 & 3 \\ 1 & 0 & 2 \end{vmatrix} = 1 \begin{vmatrix} -1 & 3 \\ 0 & 2 \end{vmatrix} - 0 \begin{vmatrix} 4 & 3 \\ 1 & 2 \end{vmatrix} + 2 \begin{vmatrix} 4 & -1 \\ 1 & 0 \end{vmatrix}$$

$$= 1(-2-0) - 0(8-3) + 2(0 + 1) = 0$$

We defined the determinant as a scalar, derived from algebraic operations on scalar entries in a matrix. However, if we put three *vectors* in the first row of a 3 × 3 matrix, then the definition still makes sense, since we would be performing scalar multiplication on those three vectors (they would be multiplied by the 2 × 2 scalar determinants as before).

This gives us a determinant that is now a vector, and lets us write the cross product of $v = v_1 i + v_2 j + v_3 k$ and $w = w_1 i + w_2 j + w_3 k$ as a determinant:

$$v \times w = \begin{vmatrix} i & j & k \\ v_1 & v_2 & v_3 \\ w_1 & w_2 & w_3 \end{vmatrix} = \begin{vmatrix} v_2 & v_3 \\ w_2 & w_3 \end{vmatrix} i - \begin{vmatrix} v_1 & v_3 \\ w_1 & w_3 \end{vmatrix} j + \begin{vmatrix} v_1 & v_2 \\ w_1 & w_2 \end{vmatrix} k$$

$= (v_2w_3 - v_3w_2)i + (v_3w_1 - v_1w_3)j + (v_1w_2 - v_2w_1)k$

Example: Let $v = 4\,i - j + 3k$ and $w = i + 2k$. Then

$$v \times w = \begin{vmatrix} i & j & k \\ 4 & -1 & 3 \\ 1 & 0 & 2 \end{vmatrix} = \begin{vmatrix} -1 & 3 \\ 0 & 2 \end{vmatrix} i - \begin{vmatrix} 4 & 3 \\ 1 & 2 \end{vmatrix} j + \begin{vmatrix} 4 & -1 \\ 1 & 0 \end{vmatrix} k = -2\,i - 5\,j + k$$

The scalar triple product can also be written as a determinant. In fact, by Example, the following theorem provides an alternate definition of the determinant of a 3 × 3 matrix as the volume (or negative volume) of a parallelepiped whose adjacent sides are the rows of the matrix. The proof is left as an exercise for the reader.

Theorem: For any vectors $u = (u_1, u_2, u_3)$, $v = (v_1, v_2, v_3)$, $w = (w_1, w_2, w_3)$ in $\mathbb{R}^3$:

$$u.(v \times w) = \begin{vmatrix} u_1 & u_2 & u_3 \\ v_1 & v_2 & v_3 \\ w_1 & w_2 & w_3 \end{vmatrix}$$

Example: Find the volume of the parallelepiped with adjacent sides $u = (2, 1, 3)$, $v = (-1, 3, 2)$, $w = (1, 1, -2)$.

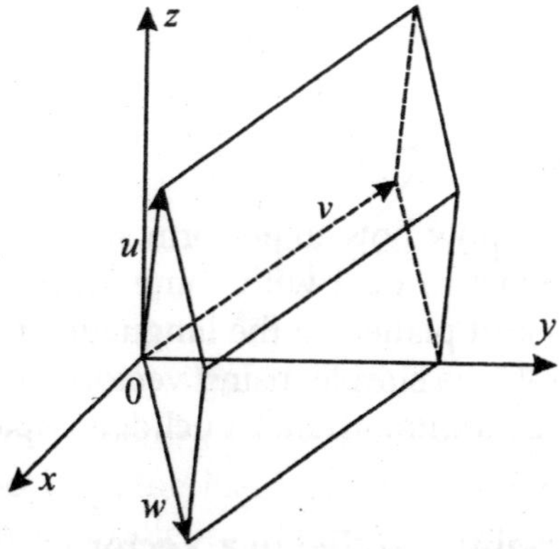

Solution: By Theorem, the volume of the parallelepiped P is the absolute value of the scalar triple product of the three adjacent sides (in any order). By Theorem,

$$u.(v \times w) = \begin{vmatrix} 2 & 1 & 3 \\ -1 & 3 & 2 \\ 1 & 1 & -2 \end{vmatrix}$$

$$= \begin{vmatrix} 3 & 2 \\ 1 & -2 \end{vmatrix} - 1\begin{vmatrix} -1 & 2 \\ 1 & -2 \end{vmatrix} + 3\begin{vmatrix} -1 & 3 \\ 1 & 1 \end{vmatrix}$$

$= 2(-8)-1(0) + 3(-4) = -28$, so

$vol\,(P) = |-28| = 28.$

Interchanging the dot and cross products can be useful in proving vector identities:

Example: Prove:

$$(u \times v) \,.\, (w \times z) = \begin{vmatrix} u.w & u.z \\ v.w & v.z \end{vmatrix}$$

for all vectors u, v, w, z in $\mathbb{R}^3$.

Solution: Let $x = u \times v$. Then

$(u \times v) \,.\, (w \times z) = x \,.\, (w \times z)$

$= w \,.\, (z \times x)$ (by formula)

$= w \,.\, (z \times (u \times v))$

$= w \,.\, ((z \,.\, v)u-(z \,.\, u)v)$ (by Theorem)

$= (z \,.\, v)(w \,.\, u)-(z \,.\, u)(w \,.\, v)$

$= (u \,.\, w)(v \,.\, z)-(u \,.\, z)(v \,.\, w)$ (by commutativity of the dot product).

$$= \begin{vmatrix} u.w & u.z \\ v.w & v.z \end{vmatrix}$$

Lines and Planes

Now that we know how to perform some operations on vectors, we can start to deal with some familiar geometric objects, like lines and planes, in the language of vectors. The reason for doing this is simple: using vectors makes it easier to study objects in 3-dimensional Euclidean space. We will first consider lines.

Line through a point, Parallel to a Vector

Let $P = (x_0, y_0, z_0)$ be a point in $\mathbb{R}^3$, let $v = (a, b, c)$ be a nonzero vector, and let L be the line through P which is parallel to v .

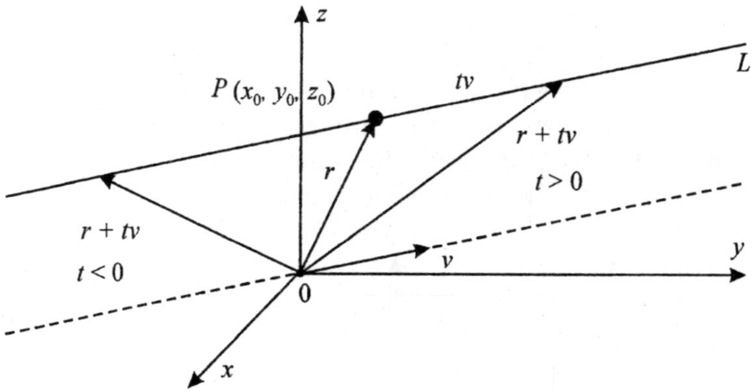

Let $r = (x_0, y_0, z_0)$ be the *vector* pointing from the origin to P. Since multiplying the vector v by a scalar t lengthens or shrinks v while preserving its direction if $t > 0$, and reversing its direction if $t < 0$, then we see from Figure that every point on the line L can be obtained by adding the vector tv to the vector r for some scalar t.

That is, as t varies over all real numbers, the vector $r + tv$ will point to every point on L. We can summarize the *vector representation of* L as follows:

For a point $P = (x_0, y_0, z_0)$ and nonzero vector v in $\mathbb{R}^3$, the line L through P parallel to v is given by,4

$$r + tv, \text{ for } -\infty < t < \infty$$

where $r = (x_0, y_0, z_0)$ is the vector pointing to P.

Note that we used the correspondence between a vector and its terminal point. Since $v = (a, b, c)$, then the terminal point of the vector $r + tv$ is $(x_0 + at, y_0 + bt, z_0 + ct)$. We then get the *parametric representation of* L with the *parameter* t:

For a point $P = (x_0, y_0, z_0)$ and nonzero vector $v = (a, b, c)$ in $\mathbb{R}^3$, the line L through P parallel to v consists of all points (x, y, z) given by

$$x = x_0 + at,\ y = y_0 + bt,\ z = z_0 + ct, \text{ for } -\infty < t < \infty$$

Note that in both representations we get the point P on L by letting $t = 0$.

In formula, if $a \neq 0$, then we can solve for the parameter t : $t = (x - x_0)/a$. We can also solve for t in terms of y and in terms of z if neither b nor c, respectively, is zero: $t = (y - y_0)/b$ and $t = (z - z_0)/c$. These three values all equal the same value t, so we can write the following system of equalities, called the *symmetric representation of L:*

For a point $P = (x_0, y_0, z_0)$ and vector $v = (a, b, c)$ in $\mathbb{R}^3$ with a, b and c all nonzero, the line L through P parallel to v consists of all points (x, y, z) given by the equations

$$\frac{x - x_0}{a} = \frac{y - y_0}{b} = \frac{z - z_0}{c}$$

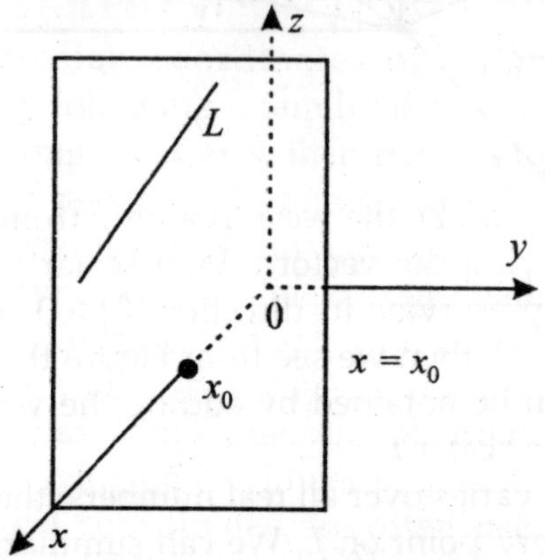

What if, say, $a = 0$ in the above scenario? We can not divide by zero, but we do know that $x = x_0 + at$, and so $x = x_0 + 0t = x_0$. Then the symmetric representation of L would be:

$$x = x_0, \frac{y - y_0}{b} = \frac{z - z_0}{c}$$

Note that this says that the line L lies in the *plane* $x = x_0$, which is parallel to the yz-plane. Similar equations can be derived for the cases when $b = 0$ or $c = 0$. You may have noticed that the vector representation of L in formula is more compact than the parametric and symmetric formulas. That is an advantage of using vector notation. Technically, though, the vector representation gives us the *vectors* whose terminal points make up the line L, not just L itself. So you have to remember to identify the vectors $r + tv$ with their terminal points. On the other hand, the parametric representation *always* gives just the points on L and nothing else.

Example : Write the line L through the point $P = (2, 3, 5)$ and parallel to the vector $v = (4, -1, 6)$, in the following forms:

(a) Vector,
(b) Parametric,
(c) Symmetric.
(d) Find two points on L distinct from P.

Solution:

(a) Let $r = (2, 3, 5)$. Then by formula (1.16), L is given by:
$r + tv = (2, 3, 5) + t\,(4, -1, 6)$, for $-\infty < t < \infty$

(b) L consists of the points (x, y, z) such that

$$x = 2 + 4t,\ y = 3-t,\ z = 5 + 6t, \text{ for } -\infty < t < \infty$$

(c) L consists of the points (x, y, z) such that

$$\frac{x-2}{4} = \frac{y-3}{-1} = \frac{z-5}{6}$$

(d) Letting $t = 1$ and $t = 2$ in part (b) yields the points (6, 2, 11) and (10, 1, 17) on L.

Line through two Points

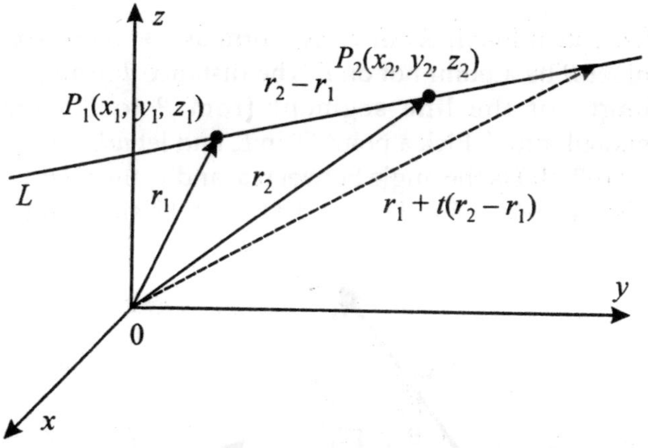

Let $P_1 = (x_1, y_1, z_1)$ and $P_2 = (x_2, y_2, z_2)$ be distinct points in $\mathbb{R}^3$, and let L be the line through P_1 and P_2. Let $r_1 = (x_1, y_1, z_1)$ and $r_2 = (x_2, y_2, z_2)$ be the vectors pointing to P_1 and P_2, respectively. Then as we can see from Figure, $r_2 - r_1$ is the vector from P_1 to P_2.

So if we multiply the vector $r_2 - r_1$ by a scalar t and add it to the vector r_1, we will get the entire line L as t varies over all real numbers. The following is a summary of the vector, parametric, and symmetric forms for the line L:

Let $P_1 = (x_1, y_1, z_1)$, $P_2 = (x_2, y_2, z_2)$ be distinct points in $\mathbb{R}^3$, and let $r_1 = (x_1, y_1, z_1)$, $r_2 = (x_2, y_2, z_2)$. Then the line L through P_1 and P_2 has the following representations:

Vector: $\quad r_1 + t(r_2 - r_1)$, for $-\infty < t < \infty$

Parametric:

$$x = x_1 + (x_2 - x_1)t,\ y = y_1 + (y_2 - y_1)t,\ z = z_1 + (z_2 - z_1)t, \quad \text{for } -\infty < t < \infty$$

Symmetric:

$$\frac{x - x_1}{x_2 - x_1} = \frac{y - y_1}{y_2 - y_1} = \frac{z - z_1}{z_2 - z_1} \quad (\text{if } x_1 \neq x_2,\ y_1 \neq y_2, \text{ and } z_1 \neq z_2)$$

Example: Write the line L through the points $P_1 = (-3, 1, -4)$ and $P_2 = (4, 4, -6)$ in parametric form.

Solution: By formula, L consists of the points (x, y, z) such that

$$x = -3 + 7t,\ y = 1 + 3t,\ z = -4 - 2t, \text{ for } -\infty < t < \infty$$

Distance between a Point and a Line

Let L be a line in $\mathbb{R}^3$ in vector form as $r + tv$ (for $-\infty < t < \infty$), and let P be a point not on L. The distance d from P to L is the length of the line segment from P to L which is perpendicular to L. Pick a point Q on L, and let w be the vector from Q to P. If θ is the angle between w and v, then $d = \|w\| \sin\theta$. So since $\|v \times w\| = \|v\|\,\|w\| \sin\theta$ and $v \neq 0$, then:

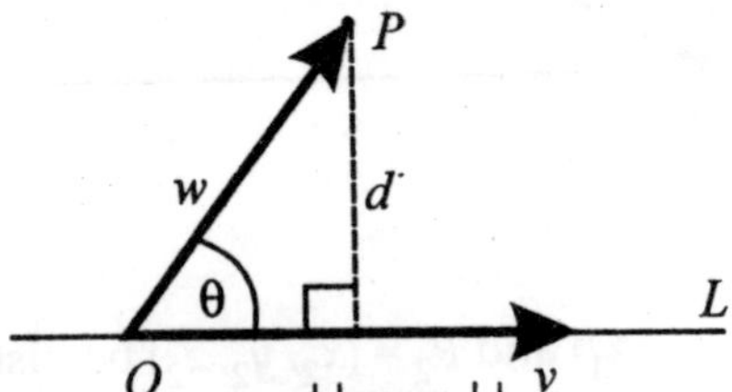

$$d = \frac{\|v \times w\|}{\|v\|}$$

Example: Find the point of intersection (if any) of the following lines:

$$\frac{x+1}{3} = \frac{y-2}{2} = \frac{z-1}{-1} \text{ and } x + 3 = \frac{y-8}{-3} = \frac{z+3}{2}$$

Solution: First we write the lines in parametric form, with parameters s and t:

$$x = -1 + 3s,\ y = 2 + 2s,$$
$$z = 1-s \text{ and } x = -3 + t,\ y = 8-3t,\ z = -3 + 2t$$

The lines intersect when $(-1 + 3s, 2 + 2s, 1-s)$
$$= (-3 + t, 8-3t, -3 + 2t) \text{ for some } s, t:$$
$$-1 + 3s = -3 + t : \Rightarrow t = 2 + 3s$$
$$2 + 2s = 8-3t : \Rightarrow 2 + 2s = 8-3(2 + 3s) = 2-9s$$
$$\Rightarrow 2s = -9s \Rightarrow s = 0 \Rightarrow t = 2 + 3(0) = 2$$
$$1-s = -3 + 2t : 1-0 = -3 + 2(2)$$
$$\Rightarrow 1 = 1 \text{ (Note that we had to check this.)}$$

Letting $s = 0$ in the equations for the first line, or letting $t = 2$ in the equations for the second line, gives the point of intersection $(-1, 2, 1)$. We will now consider planes in 3-dimensional Euclidean space.

Plane through a Point, Perpendicular to a Vector

Let P be a plane in $\mathbb{R}^3$, and suppose it contains a point $P_0 = (x_0, y_0, z_0)$. Let $n = (a, b, c)$ be a nonzero vector which is perpendicular to the plane P. Such a vector is called a normal vector (or just a *normal*) to the plane. Now let (x, y, z) be any point in the plane P. Then the vector $r = (x-x_0, y-y_0, z-z_0)$ lies in the plane P. So if $r \neq 0$, then $r \perp n$ and hence $n \,.\, r = 0$. And if $r = 0$ then we still have $n \,.\, r = 0$.

Conversely, if (x, y, z) is any point in $\mathbb{R}^3$ such that $r = (x-x_0, y-y_0, z-z_0) \neq 0$ and $n \,.\, r = 0$, then $r \perp n$ and so (x, y, z) lies in P. This proves the following theorem:

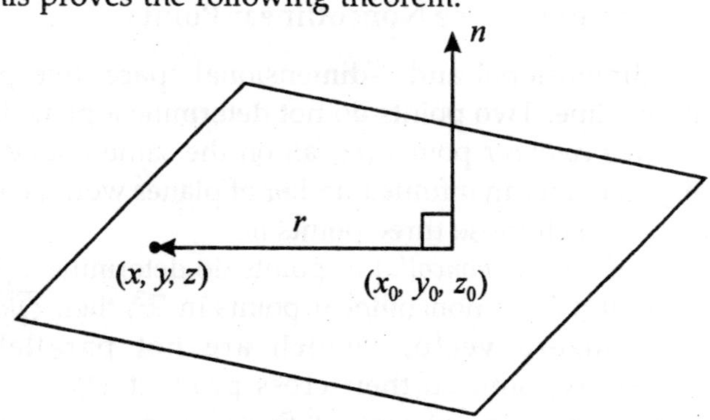

Fig. The Plane P

Theorem: Let P be a plane in $\mathbb{R}^3$, let (x_0, y_0, z_0) be a point in P, and let n = (a, b, c) be a nonzero vector whic is perpendicular to P.

Then P consists of the points (x, y, z) satisfying the vector equation:

$$n \cdot r = 0$$

where,

$$r = (x-x_0, y-y_0, z-z_0),$$

or equivalently:

$$a(x-x_0) + b(y-y_0) + c(z-z_0) = 0$$

The above equation is called the point-normal form of the plane P.

Example: Find the equation of the plane P containing the point (–3, 1, 3) and perpendicular to the vector $n = (2, 4, 8)$.

Solution: By formula, the plane P consists of all points (x, y, z) such that:

$$2(x + 3) + 4(y-1) + 8(z-3) = 0$$

If we multiply out the terms in formula and combine the constant terms, we get an equation of the plane in normal form:

$$ax + by + cz + d = 0$$

For example, the normal form of the plane in Example is ,

$$2x + 4y + 8z-22 = 0.$$

Plane Containing three Noncollinear Points

In 2-dimensional and 3-dimensional space, two points determine a line. Two points do not determine a plane in $\mathbb{R}^3$. In fact, three *collinear* points (*i.e.* all on the same line) do not determine a plane; an infinite number of planes would contain the line on which those three points lie.

However, three *noncollinear* points do determine a plane. For if Q, R and S are noncollinear points in $\mathbb{R}^3$, then $\overrightarrow{QR}$ and $\overrightarrow{QS}$ are nonzero vectors which are not parallel (by noncollinearity), and so their cross product $\overrightarrow{QR} \times \overrightarrow{QS}$ is perpendicular to both $\overrightarrow{QR}$ and $\overrightarrow{QS}$ So $\overrightarrow{QR}$ and $\overrightarrow{QS}$ (and hence Q, R and S) lie in the plane through the point Q with normal vector $n = \overrightarrow{QR} \times \overrightarrow{QS}$.

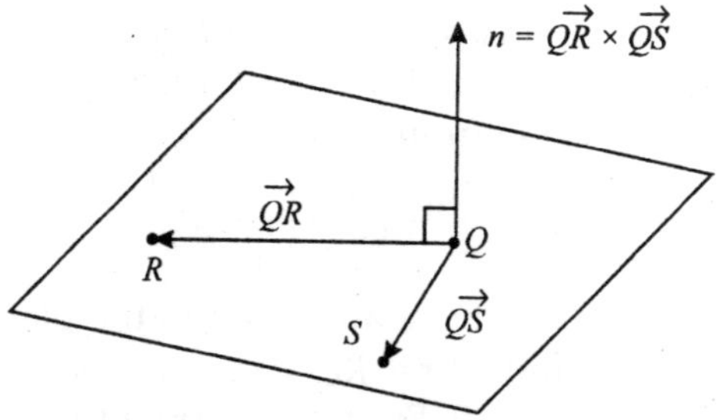

Fig. Noncollinear points Q, R, S

Example: Find the equation of the plane P containing the points (2, 1, 3), (1, –1, 2) and (3, 2, 1).

Solution: Let,

$Q = (2, 1, 3)$,

$R = (1, -1, 2)$ and

$S = (3, 2, 1)$.

Then for the vectors $\overrightarrow{QR} = (-1, -2, -1)$ and $\overrightarrow{QS} = (1, 1, -2)$, the plane P has a normal vector

$$n = \overrightarrow{QR} \times \overrightarrow{QS} = (-1, -2, -1) \times (1, 1, -2) = (5, -3, 1)$$

So using formula with the point Q (we could also use R or S), the plane P consists of all points (x, y, z) such that:

$$5(x-2)-3(y-1) + (z-3) = 0$$

or in normal form,

$$5x-3y + z-10 = 0$$

We mentioned earlier that skew lines in $\mathbb{R}^3$ lie on separate, parallel planes. So two skew lines do not determine a plane. But two (nonidentical) lines which either intersect or are parallel do determine a plane. In both cases, to find the equation of the plane that contains those two lines, simply pick from the two lines a total of three noncollinear points (*i.e.* one point from one line and two points from the other), then use the technique above, to write the equation. We will leave examples of this as exercises for the reader.

Distance between a Point and a Plane

The distance between a point in $\mathbb{R}^3$ and a plane is the length of the line segment from that point to the plane which is perpendicular to the plane. The following theorem gives a formula for that distance.

Theorem: Let $Q = (x_0, y_0, z_0)$ be a point in $\mathbb{R}^3$, and let P be a plane with normal form $ax + by + cz + d = 0$ that does not contain Q. Then the distance D from Q to P is:

$$D = \frac{|ax_0 + by_0 + cz_0 + d|}{\sqrt{a^2 + b^2 + c^2}}$$

Proof: Let $R = (x, y, z)$ be any point in the plane P (so that $ax + by + cz + d = 0$) and let $r = \overrightarrow{RQ} = (x_0 - x, y_0 - y, z_0 - z)$. Then $r \neq 0$ since Q does not lie in P. From the normal form equation for P, we know that $n = (a, b, c)$ is a normal vector for P. Now, any plane divides $\mathbb{R}^3$ into two disjoint parts. Assume that n points towards the side of P where the point Q is located. Place n so that its initial point is at R, and let θ be the angle between r and n. Then $0° < \theta < 90°$, so $\cos\theta > 0$. Thus, the distance D is $\cos\theta \, \|r\| = |\cos\theta| \, \|r\|$.

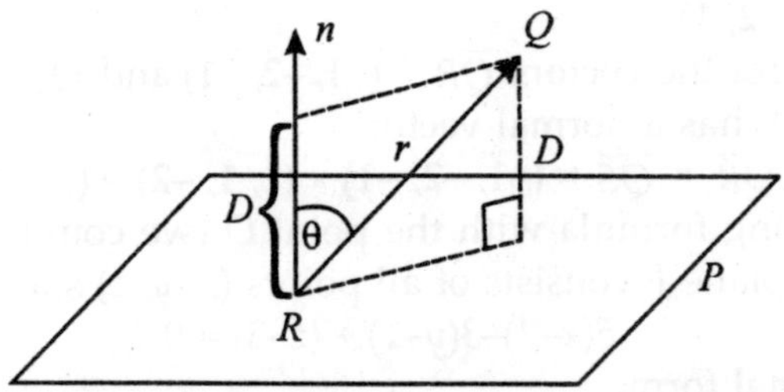

By Theorem, we know that $\cos\theta = \dfrac{n.r}{\|n\|\|r\|}$, so

$$D = |\cos\theta|\,\|r\| = \frac{|n.r|}{\|n\|\|r\|}\|r\| = \frac{|n.r|}{\|n\|}$$

$$= \frac{|a(x_0 - x) + b(y_0 - y) + c(z_0 - z)|}{\sqrt{a^2 + b^2 + c^2}}$$

$$= \frac{|ax_0 + by_0 + cz_0 - (ax + by + cz)|}{\sqrt{a^2 + b^2 + c^2}}$$

$$= \frac{|ax_0 + by_0 + cz_0 - (-d)|}{\sqrt{a^2 + b^2 + c^2}} = \frac{|ax_0 + by_0 + cz_0 + d|}{\sqrt{a^2 + b^2 + c^2}}$$

If n points away from the side of P where the point Q is located, then $90° < \theta < 180°$ and so $\cos\theta < 0$.

The distance D is then $|\cos\theta| \, \|r\|$, and thus repeating the same argument as above still gives the same result.

Example: Find the distance D from (2, 4, –5) to the plane from Example.

Solution: Recall that the plane is given by

$5x-3y + z-10 = 0$. So

$$D = \frac{|5(2)-3(4)+1(-5)-10|}{\sqrt{5^2+(-3)^2+1^2}} = \frac{|-17|}{\sqrt{35}} = \frac{17}{\sqrt{35}} \approx 2.87$$

Line of Intersection of two Planes

Note that two planes are parallel if they have normal vectors that are parallel, and the planes are perpendicular if their normal vectors are perpendicular. If two planes do intersect, they do so in a line. Suppose that two planes P_1 and P_2 with normal vectors n_1 and n_2, respectívely, intersect in a line L.

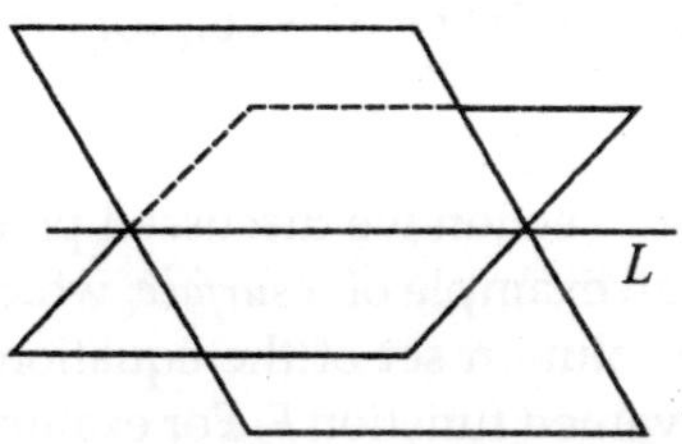

Since $n_1 \times n_2 \perp n_1$, then $n_1 \times n_2$ is parallel to the plane P_1. Likewise, $n_1 \times n_2 \perp n_2$ means that $n_1 \times n_2$ is also parallel to P_2. Thus, $n_1 \times n_2$ is parallel to the intersection of P_1 and P_2, *i.e.* $n_1 \times n_2$ is parallel to L. Thus, we can write L in the following vector form:

$$L : r + t(n_1 \times n_2), \text{ for } -\infty < t < \infty$$

where r is any vector pointing to a point belonging to both planes. To find a point in both planes, find a common solution (x, y, z) to the two normal form equations of the planes. This can often be made easier by setting one of the coordinate variables to zero, which leaves you to solve two equations in just two unknowns.

Example: Find the line of intersection L of the planes $5x-3y + z-10 = 0$ and $2x + 4y-z + 3 = 0$.

Solution: The plane $5x-3y + z$ " $10 = 0$ has normal vector $n_1 = (5, -3, 1)$ and the plane $2x + 4y-z + 3 = 0$ has normal vector $n_2 = (2, 4, -1)$.

Since n_1 and n_2 are not scalar multiples, then the two planes are not parallel and hence will intersect. A point (x, y, z) on both planes will satisfy the following system of two equations in three unknowns:

$$5x-3y+z-10=0$$
$$2x+4y-z+3=0$$

Set $x = 0$ (why is that a good choice?). Then the above equations are reduced to:

$$-3y+z-10=0$$
$$4y-z+3=0$$

The second equation gives $z = 4y + 3$, substituting that into the first equation gives $y = 7$. Then $z = 31$, and so the point (0, 7, 31) is on L. Since $n_1 \times n_2 = (-1, 7, 26)$, then L is given by:

$$\underline{r} + t(n_1 \times n_2) = (0, 7, 31) + t(-1, 7, 26), \text{ for } -\infty < t < \infty$$

or in parametric form:

$$x = -t,\ y = 7 + 7t,\ z = 31 + 26t, \text{ for } -\infty < t < \infty$$

Surfaces

In the previous section we discussed planes in Euclidean space. A plane is an example of a *surface*, which we will define informally as the solution set of the equation $F(x, y, z) = 0$ in $\mathbb{R}^3$, for some real-valued function F. For example, a plane given by $ax + by + cz + d = 0$ is the solution set of $F(x, y, z) = 0$ for the function $F(x, y, z) = ax + by + cz + d$. Surfaces are 2-dimensional. The plane is the simplest surface, since it is "flat". In this section we will look at some surfaces that are more complex, the most important of which are the sphere and the cylinder.

Definition: A sphere S is the set of all points (x, y, z) in $\mathbb{R}^3$ which are a fixed distance r (called the radius) from a fixed point $P_0 = (x_0, y_0, z_0)$ (called the centre of the sphere):

$$S = \{(x, y, z) : (x-x_0)^2 + (y-y_0)^2 + (z-z_0)^2 = r^2\}$$

Using vector notation, this can be written in the equivalent form:

$$S = \{x : \|\, x-x_0 \,\| = r\}$$

where $x = (x, y, z)$ and $x_0 = (x_0, y_0, z_0)$ are vectors.

Figure illustrates the vectorial approach to spheres. Note in Figure that the intersection of the sphere with the xy-plane is a circle of radius r (*i.e.* a *great circle*, given by $x^2 + y^2 = r^2$ as a subset of $\mathbb{R}^2$). Similarly for the intersections with the xz-plane and the yz-plane. In general, a plane intersects a sphere either at a single point or in a circle.

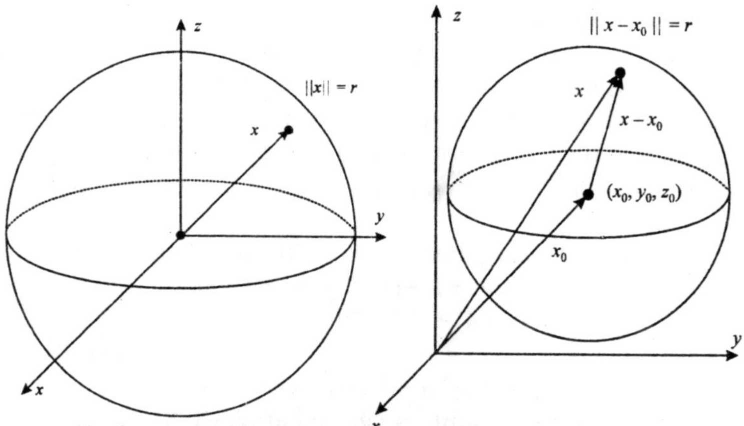

(a) radius r, center (0, 0, 0) (b) radius r, center (x_0, y_0, z_0)

Fig. Spheres in $\mathbb{R}^3$

Example : Find the intersection of the sphere $x^2 + y^2 + z^2 = 169$ with the plane $z = 12$.

Solution: The sphere is centred at the origin and has radius $13 = \sqrt{169}$, so it does intersect the plane $z = 12$.

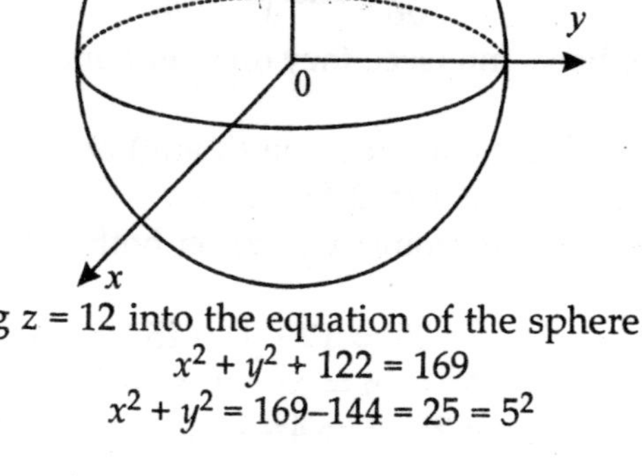

Putting $z = 12$ into the equation of the sphere gives

$$x^2 + y^2 + 122 = 169$$

$$x^2 + y^2 = 169–144 = 25 = 5^2$$

which is a circle of radius 5 centred at (0, 0, 12), parallel to the xy-plane.

If the equation in formula is multiplied out, we get an equation of the form:

$$x^2 + y^2 + z^2 + ax + by + cz + d = 0$$

for some constants a, b, c and d. Conversely, an equation of this form *may* describe a sphere, which can be determined by completing the square for the x, y and z variables.

Example: Is $2x^2 + 2y^2 + 2z^2 - 8x + 4y - 16z + 10 = 0$ the equation of a sphere?

Solution: Dividing both sides of the equation by 2 gives

$$x^2 + y^2 + z^2 - 4x + 2y - 8z + 5 = 0$$

$$(x^2 - 4x + 4) + (y^2 + 2y + 1) + (z^2 - 8z + 16) + 5 - 4 - 1 - 16 = 0$$

$$(x-2)^2 + (y + 1)2 + (z-4)^2 = 16$$

which is a sphere of radius 4 centred at (2, –1, 4).

Example: Find the points(s) of intersection (if any) of the sphere from Example and the line $x = 3 + t$, $y = 1 + 2t$, $z = 3-t$.

Solution: Put the equations of the line into the equation of the sphere, which was $(x-2)^2 + (y + 1)^2 + (z-4)^2 = 16$, and solve for t:

$$(3 + t-2)^2 + (1 + 2t + 1)^2 + (3-t-4)^2 = 16$$

$$(t + 1)^2 + (2t + 2)^2 + (-t-1)^2 = 16$$

$$6t^2 + 12t - 10 = 0$$

The quadratic formula gives the solutions $t = -1 \pm \frac{4}{\sqrt{6}}$. Putting those two values into the equations of the line gives the following two points of intersection:

$$\left(2+\frac{4}{\sqrt{6}},-1+\frac{8}{\sqrt{6}},4-\frac{4}{\sqrt{6}}\right) \text{ and } \left(2-\frac{4}{\sqrt{6}},-1-\frac{8}{\sqrt{6}},4+\frac{4}{\sqrt{6}}\right)$$

If two spheres intersect, they do so either at a single point or in a circle.

Example: Find the intersection (if any) of the spheres $x^2 + y^2 + z^2 = 25$ and $x^2 + y^2 + (z-2)^2 = 16$.

Solution: For any point (x, y, z) on both spheres, we see that,

$$x^2 + y^2 + z^2 = 25 \Rightarrow x2 + y2 = 25 \text{ “ } z2, \text{ and}$$

$$x^2 + y^2 + (z-2)^2 = 16 \Rightarrow x^2 + y^2 = 16-(z-2)^2, \text{ so}$$

$$16-(z-2)^2 = 25-z^2 \Rightarrow 4z-4 = 9 \Rightarrow z = 13/4$$

$$\Rightarrow x^2 + y^2 = 25-(13/4)^2 = 231/16$$

$\therefore$ The intersection is the circle $x^2 + y^2 = \dfrac{231}{16}$ of radius $\dfrac{\sqrt{231}}{4} \approx 3.8$ centred at $\left(0,0,\dfrac{13}{4}\right)$.

The cylinders that we will consider are *right circular cylinders*. These are cylinders obtained by moving a line L along a circle C in $\mathbb{R}^3$ in a way so that L is always perpendicular to the plane containing C. We will only consider the cases where the plane containing C is parallel to one of the three coordinate planes.

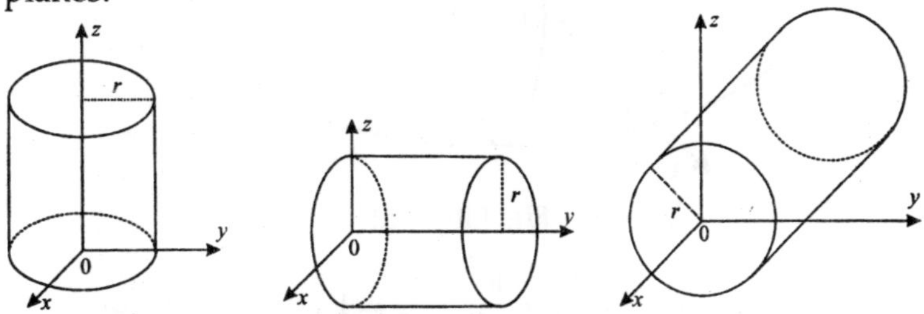

(a) $x^2 + y^2 = r^2$, any z
(b) $x^2 + z^2 = r^2$, any y
(c) $y^2 + z^2 = r^2$, any x

Fig. Cylinders in $\mathbb{R}^3$

For example, the equation of a cylinder whose base circle C lies in the xy-plane and is centred at $(a, b, 0)$ and has radius r is

$$(x-a)^2 + (y-b)^2 = r^2,$$

where the value of the z coordinate is unrestricted. Similar equations can be written when the base circle lies in one of the other coordinate planes. A plane intersects a right circular cylinder in a circle, ellipse, or one or two lines, depending on whether that plane is parallel, oblique, or perpendicular, respectively, to the plane containing C.

The intersection of a surface with a plane is called the trace of the surface. The equations of spheres and cylinders are examples of *second-degree equations* in $\mathbb{R}^3$, *i.e.* equations of the form $Ax^2 + By^2 + Cz^2 + Dxy + Exz + Fyz + Gx + Hy + Iz + J = 0$ for some constants $A, B, ..., J$. If the above equation is not that of a sphere, cylinder, plane, line or point, then the resulting surface is called a quadric surface.

One type of quadric surface is the ellipsoid, given by an equation of the form:

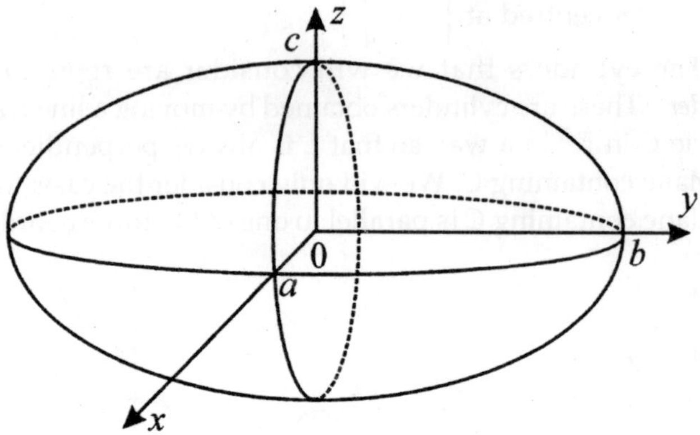

Fig. Ellipsoid

$$\frac{x^2}{a^2}+\frac{y^2}{b^2}+\frac{z^2}{c^2}=1$$

In the case where $a = b = c$, this is just a sphere. In general, an ellipsoid is egg-shaped (think of an ellipse rotated around its major axis). Its traces in the coordinate planes are ellipses.

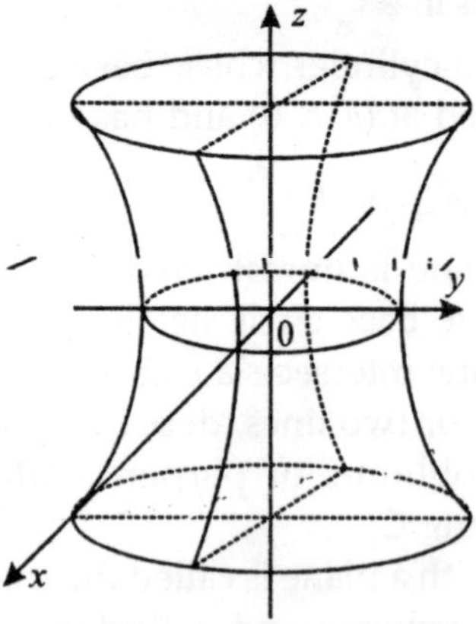

Fig. Hyperboloid of one sheet

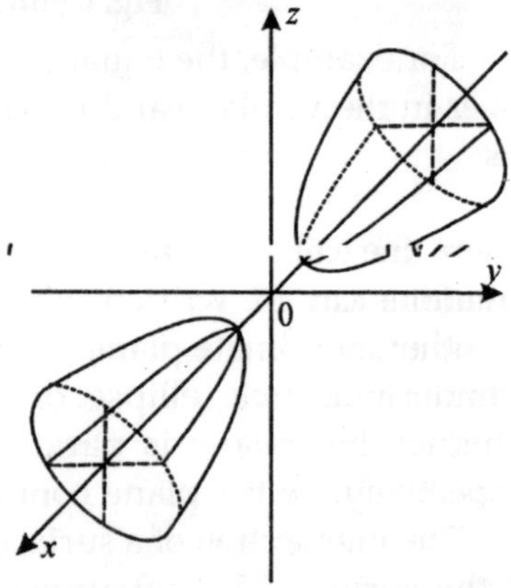

Fig. Hyperboloid of two sheets

Two other types of quadric surfaces are the hyperboloid of one sheet, given by an equation of the form:

$$\frac{x^2}{a^2}+\frac{y^2}{b^2}-\frac{z^2}{c^2}=1$$

and the hyperboloid of two sheets, whose equation has the form:

$$\frac{x^2}{a^2}-\frac{y^2}{b^2}-\frac{z^2}{c^2}=1$$

For the hyperboloid of one sheet, the trace in any plane parallel to the xy-plane is an ellipse. The traces in the planes parallel to the xz- or yz-planes are hyperbolas, except for the special cases $x = \pm a$ and $y = \pm b$; in those planes the traces are pairs of intersecting lines. For the hyperboloid of two sheets, the trace in any plane parallel to the xy- or xzplane is a hyperbola. There is no trace in the yz-plane. In any plane parallel to the yz-plane for which $|x| > |a|$, the trace is an ellipse. The elliptic paraboloid is another type of quadric surface, whose equation has the form:

$$\frac{x^2}{a^2}+\frac{y^2}{b^2}=\frac{z}{c}$$

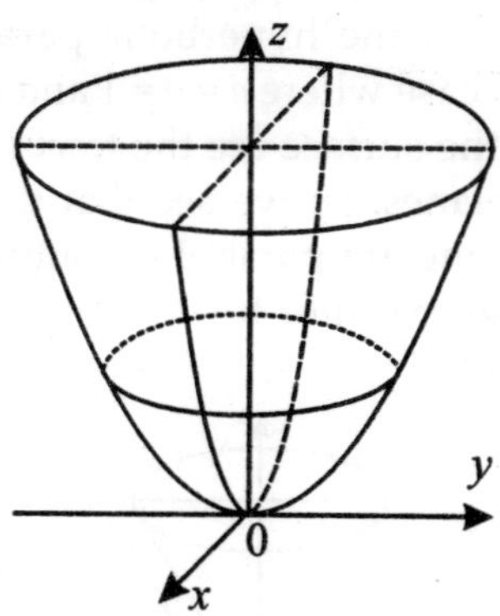

Fig. Paraboloid

The traces in planes parallel to the xy-plane are ellipses, though in the xy-plane itself the trace is a single point. The traces in planes parallel to the xz- or yz-planes are parabolas. Figure shows the case where $c > 0$. When $c < 0$ the surface is turned downward. In the case where $a = b$, the surface is called a *paraboloid of revolution,* which is often used as a reflecting surface, *e.g.* in vehicle headlights. A more complicated quadric surface is the hyperbolic paraboloid, given by:

$$\frac{x^2}{a^2} - \frac{y^2}{b^2} = \frac{z}{c}$$

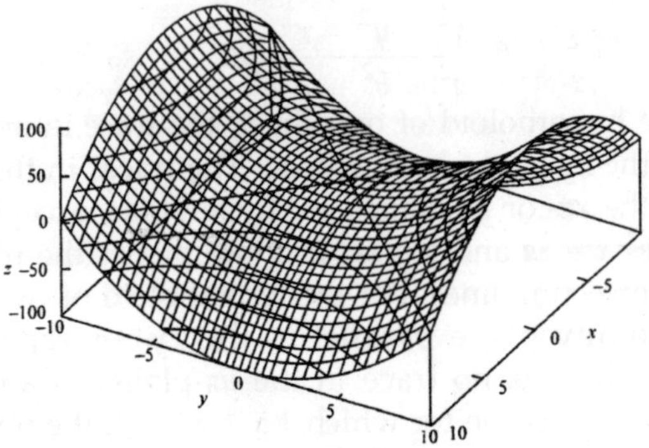

Fig. Hyperbolic Paraboloid

The hyperbolic paraboloid can be tricky to draw; using graphing software on a computer can make it easier. For example, Figure was created using the free Gnuplot package. It shows the graph of the hyperbolic paraboloid $z = y^2 - x^2$, which is the special case where $a = b = 1$ and $c = -1$ in equation. The mesh lines on the surface are the traces in planes parallel to the coordinate planes. So we see that the traces in planes parallel to the xz-plane are parabolas pointing upward, while the traces in planes parallel to the yz-plane are parabolas pointing downward.

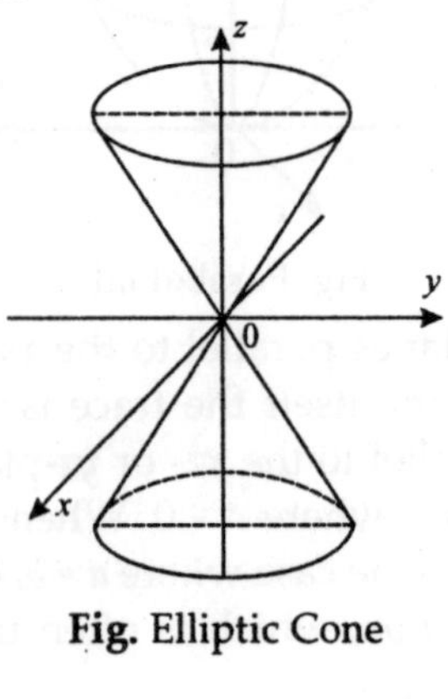

Fig. Elliptic Cone

Also, notice that the traces in planes parallel to the xy-plane are hyperbolas, though in the xy-plane itself the trace is a pair of intersecting lines through the origin.

This is true in general when $c < 0$ in equation. When $c > 0$, the surface would be similar to that in Figure, only rotated 90° around the z-axis and the nature of the traces in planes parallel to the xz- or yz-planes would be reversed.

The last type of quadric surface that we will consider is the elliptic cone, which has an equation of the form:

$$\frac{x^2}{a^2}+\frac{y^2}{b^2}-\frac{z^2}{c^2}=0$$

The traces in planes parallel to the xy-plane are ellipses, except in the xy-plane itself where the trace is a single point. The traces in planes parallel to the xz- or yz-planes are hyperbolas, except in the xz- and yz-planes themselves where the traces are pairs of intersecting lines.

Notice that every point on the elliptic cone is on a line which lies entirely on the surface; in Figure these lines all go through the origin. This makes the elliptic cone an example of a *ruled surface*. The cylinder is also a ruled surface.

What may not be as obvious is that both the hyperboloid of one sheet and the hyperbolic paraboloid are ruled surfaces. In fact, on both surfaces there are *two* lines through each point on the surface. Such surfaces are called *doubly ruled surfaces,* and the pairs of lines are called a *regulus*.

It is clear that for each of the six types of quadric surfaces that we discussed, the surface can be translated away from the origin (*e.g.* by replacing x^2 by $(x-x_0)^2$ in its equation). It can be proved11 that *every* quadric surface can be translated and/or rotated so that its equation matches one of the six types that we described. For example, $z = 2xy$ is a case of equation with "mixed" variables, *e.g.* with $D \neq 0$ so that we get an xy term.

This equation does not match any of the types we considered. However, by rotating the x- and y-axes by 45° in the xy-plane by means of the coordinate transformation,

$x = (x'-y')/\sqrt{2}$,

$y = (x' + y')/\sqrt{2}$,

$z = z'$,

then $z = 2xy$ becomes the hyperbolic paraboloid $z' = (x')^2-(y')^2$ in the (x', y', z') coordinate system. That is, $z = 2xy$ is a hyperbolic paraboloid as in equation, but rotated 45° in the xy-plane.

CURVILINEAR COORDINATES

The Cartesian coordinates of a point (x, y, z) are determined by following a family of straight paths from the origin: first along the x-axis, then parallel to the y-axis, then parallel to the z-axis, as in Figure. In *curvilinear coordinate systems,* these paths can be curved.

The two types of curvilinear coordinates which we will consider are cylindrical and spherical coordinates. Instead of referencing a point in terms of sides of a rectangular parallelepiped, as with Cartesian coordinates, we will think of the point as lying on a cylinder or sphere.

Cylindrical coordinates are often used when there is symmetry around the z-axis, while spherical coordinates are useful when there is symmetry about the origin.

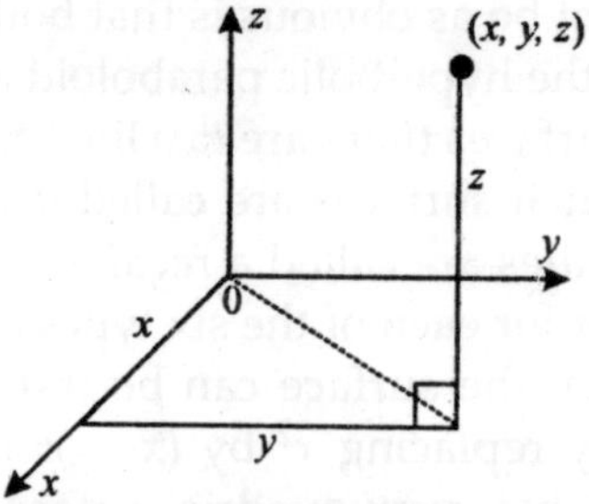

Let $P = (x, y, z)$ be a point in Cartesian coordinates in $\mathbb{R}^3$, and let $P_0 = (x, y, 0)$ be the projection of P upon the xy-plane. Treating (x, y) as a point in $\mathbb{R}^2$, let (r, θ) be its polar coordinates. Let ρ be the length of the line segment from the origin to P, and let ϕ be the angle between that line segment and the positive z-axis.

ϕ is called the *zenith angle*. Then the cylindrical coordinates (r, θ, z) and the spherical coordinates (ρ, θ, z) of $P(x, y, z)$ are defined as follows:

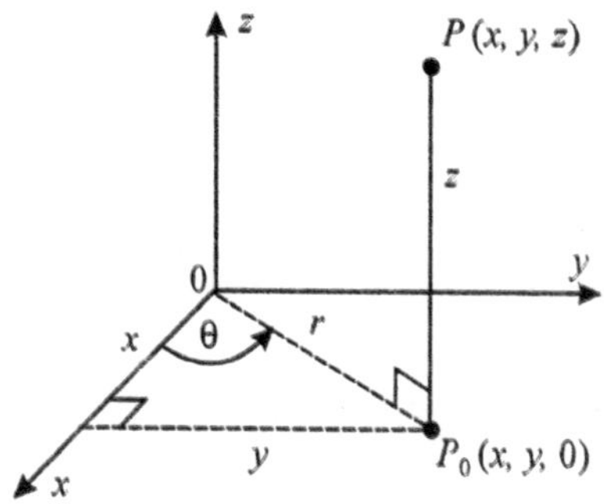

CYLINDRICAL COORDINATES

Cylindrical coordinates: (r, θ, z);

$x = r\cos\theta$ $\qquad r = \sqrt{x^2 + y^2}$

$y = r\sin\theta$ $\qquad \theta = \tan^{-1}\left(\dfrac{y}{x}\right)$

$z = z$

where $0 \le \theta \le \pi$ if $y \ge 0$ and $\pi < \theta < 2\pi$ if $y < 0$

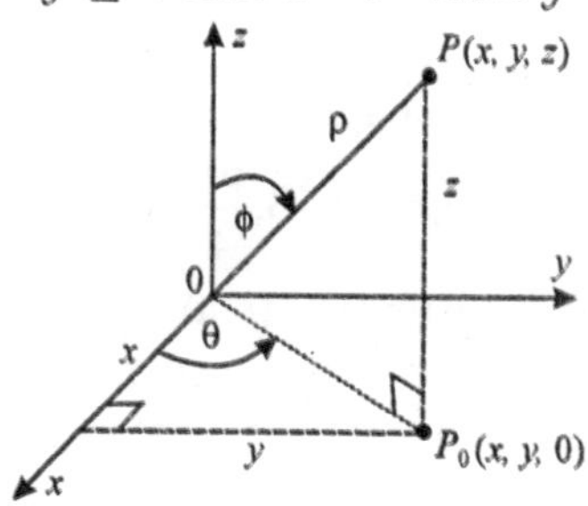

SPHERICAL COORDINATES

Spherical Coordinates (ρ, θ, ϕ):

$$x = \rho\sin\phi\cos\theta$$

$$\rho = \sqrt{x^2 + y^2 + z^2}$$

$$y = \rho\sin\phi\sin\theta$$

$$\theta = \tan^{-1}\left(\frac{y}{x}\right)$$

$$z = \rho\cos\phi$$

$$\phi = \cos^{-1}\left(\frac{z}{\sqrt{x^2 + y^2 + z^2}}\right)$$

where $0 \leq \theta \leq \pi$ if $y \geq 0$ and $\pi < \theta < 2\pi$ if $y < 0$. Both θ and ϕ aremeasured in radians. Note that $r \geq 0$, $0 \leq \theta < 2\pi$, $\rho \geq 0$ and $0 \leq \phi \leq \pi$. Also, θ is undefined when $(x, y) = (0, 0)$, and ϕ is undefined when $(x, y, z) = (0, 0, 0)$.

Example: Convert the point (–2, –2, 1) from Cartesian coordinates to (a) cylindrical and (b) spherical coordinates.

Solution:

(a) $r = \sqrt{(-2)^2 + (-2)^2} = 2\sqrt{2}, \theta = \tan^{-1}\left(\frac{-2}{-2}\right)$

$$\tan^{-1}(1) = \frac{5\pi}{4},$$

since $y = -2 < 0$.

$$\therefore\ (r, \theta, z) = \left(2\sqrt{2}, \frac{5\pi}{4}, 1\right)$$

(b) $\rho = \sqrt{(-2)^2 + (-2)^2 + 1^2} = \sqrt{9} = 3,$

$$\phi = \cos^{-1}\left(\frac{1}{3}\right) \approx 1.23 \text{ radians.}$$

$$\therefore (\rho, \theta, \phi,) = \left(3, \frac{5\pi}{4}, 1.23\right)$$

For cylindrical coordinates (r, θ, z), and constants r_0, θ_0 and z_0, we see from figure that the surface $r = r_0$ is a cylinder of radius r_0 centred along the z-axis, the surface $\theta = \theta_0$ is a half-plane emanating from the z-axis, and the surface $z = z_0$ is a plane parallel to the xy-plane.

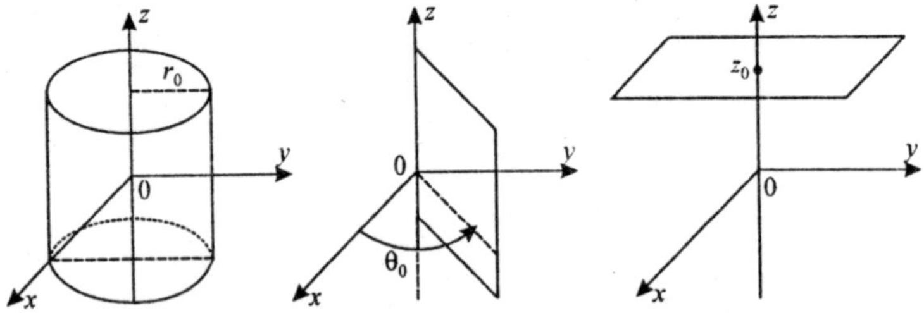

(a) $r = r_0$ (b) $\theta = \theta_0$ (c) $z = z_0$

Fig. Cylindrical Coordinate Surfaces

For spherical coordinates (ρ, θ, ϕ), and constants ρ_0, θ_0 and ϕ_0, we see from Figure that the surface $\rho = \rho_0$ is a sphere of radius ρ_0 centred at the origin, the surface $\theta = \theta_0$ is a half-plane emanating from the z-axis, and the surface $\phi = \phi_0$ is a circular cone whose vertex is at the origin.

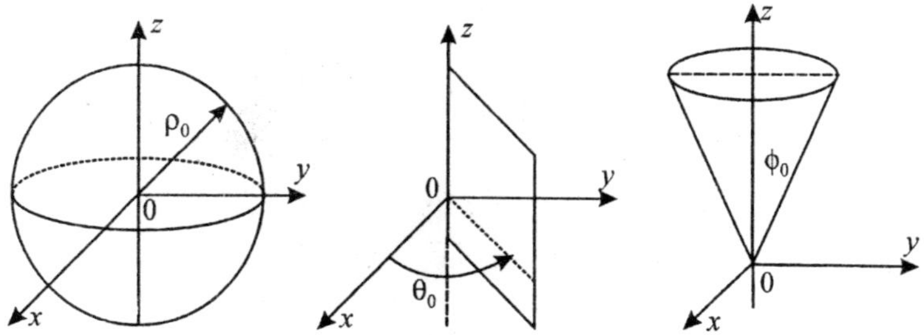

(a) $\rho = \rho_0$ (b) $\theta = \theta_0$ (c) $\phi = \phi_0$

Fig. Spherical Coordinate Surfaces

Figures show how these coordinate systems got their names. Sometimes the equation of a surface in Cartesian coordinates can be transformed into a simpler equation in some other coordinate system, as in the following example.

Example: Write the equation of the cylinder $x^2 + y^2 = 4$ in cylindrical coordinates.

Solution: Since $r = \sqrt{x^2+y^2}$, then the equation in cylindrical coordinates is $r = 2$. Using spherical coordinates to write the equation of a sphere does not necessarily make the equation simpler, if the sphere is not centred at the origin.

Example: Write the equation $(x-2)^2 + (y-1)^2 + z^2 = 9$ in spherical coordinates.

Solution: Multiplying the equation out gives,

$$x^2 + y^2 + z^2 - 4x - 2y + 5 = 9, \text{ so we get}$$

$$r^2 - 4r \sin\phi \cos\theta - 2r \sin\phi \sin\theta - 4 = 0, \text{ or}$$

$$r^2 - 2 \sin\phi\,(2\cos\theta - \sin\theta)r - 4 = 0$$

after combining terms. Note that this actually makes it more difficult to figure out what the surface is, as opposed to the Cartesian equation where you could immediately identify the surface as a sphere of radius 3 centred at (2, 1, 0).

Example: Describe the surface given by $\theta = z$ in cylindrical coordinates.

Solution: This surface is called a *helicoid*. As the (vertical) z coordinate increases, so does the angle θ, while the radius *r* is unrestricted. So this sweeps out a (ruled!) surface shaped like a spiral staircase, where the spiral has an infinite radius. Figure shows a section of this surface restricted to $0 \leq z \leq 4\pi$ and $0 \leq r \leq 2$.

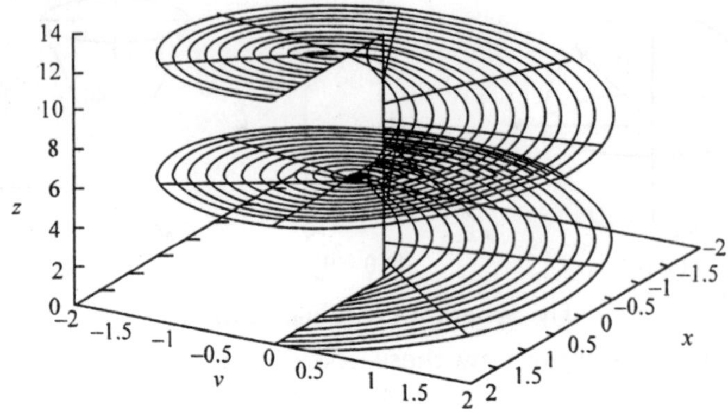

Fig. Helicoid $\theta = z$

Vector-Valued Functions

Now that we are familiar with vectors and their operations, we can begin discussing functions whose values are vectors.

Definition: A vector-valued function of a real variable is a rule that associates a vector $f(t)$ with a real number t, where t is in some subset D of $\mathbb{R}^1$ (called the domain of f). We write f: $D \to \mathbb{R}^3$ to denote that f is a mapping of D into $\mathbb{R}^3$.

For example, $f(t) = ti + t^2j + t^3k$ is a vector-valued function in $\mathbb{R}^3$, defined for all real numbers t. We would write $f: \mathbb{R} \to \mathbb{R}^3$. At $t = 1$ the value of the function is the vector $i + j + k$, which in Cartesian coordinates has the terminal point $(1, 1, 1)$.

A vector-valued function of a real variable can be written in component form as,

$$f(t) = f_1(t)i + f_2(t)j + f_3(t)k$$

or in the form,

$$f(t) = (f_1(t), f_2(t), f_3(t))$$

for some real-valued functions $f_1(t)$, $f2(t)$, $f3(t)$, called the *component functions* of f. The first form is often used when emphasizing that $f(t)$ is a vector, and the second form is useful when considering just the terminal points of the vectors. By identifying vectors with their terminal points, a curve in space can be written as a vector-valued function.

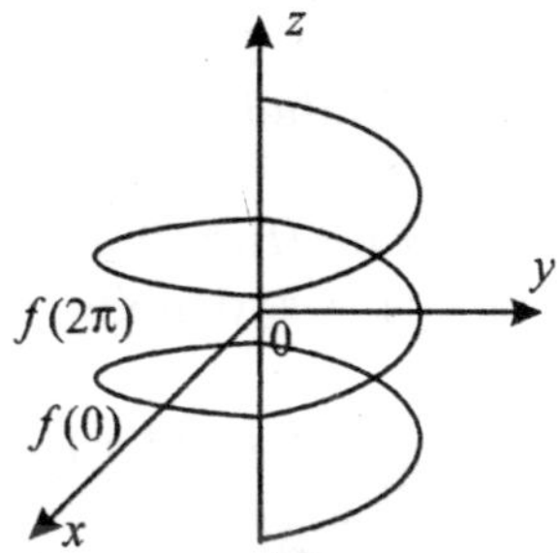

Example: Define $f: \mathbb{R} \to \mathbb{R}^3$ by $f(t) = (\cos t, \sin t, t)$. This is the equation of a *helix*. As the value of t increases, the terminal points of $f(t)$ trace out a curve spiraling upward. For each t, the x- and y-coordinates of $f(t)$ are $x = \cos t$ and $y = \sin t$, so

$$x^2 + y^2 = \cos^2 t + \sin^2 t = 1.$$

Thus, the curve lies on the surface of the right circular cylinder $x^2 + y^2 = 1$. It may help to think of vector-valued functions of a real variable in $\mathbb{R}^3$ as a generalization of the parametric functions in $\mathbb{R}^2$ which you learned about in single-variable calculus. Much of the theory of real-valued functions of a single real variable can be applied to vector-valued functions of a real variable. Since each of the three component functions are real-valued, it will sometimes be the case that results from singlevariable calculus can simply be applied to each of the component functions to yield a similar result for the vector-valued function. However, there are times when such generalizations do not hold. The concept of a limit, though, can be extended naturally to vector-valued functions, as in the following definition.

Definition: Let $f(t)$ be a vector-valued function, let a be a real number and let c be a vector. Then we say that the limit of $f(t)$ as t approaches a equals c, written as $\lim_{t \to a} f(t) = c$, if $\lim_{t \to a} \| f(t) - c \| = 0$. If $f(t) = (f_1(t), f_2(t), f_3(t))$, then,

$$\lim_{t \to a} f(t) = \left(\lim_{t \to a} f_1(t), \lim_{t \to a} f_2(t), \lim_{t \to a} f_3(t) \right)$$

provided that all three limits on the right side exist. The above definition shows that continuity and the derivative of vector-valued functions can also be defined in terms of its component functions.

Definition: Let $f(t) = (f_1(t), f_2(t), f_3(t))$ be a vector-valued function, and let a be a real number in its domain. Then $f(t)$ is continuous at a if $\lim_{t \to a} f(t) = f(a)$. Equivalently, $f(t)$ is continuous at a if and only if $f_1(t), f_2(t)$, and $f_3(t)$ are continuous at a. The derivative of $f(t)$ at a, denoted by $f'(a)$ or $\frac{df}{dt}(a)$, is the limit,

$$f'(a) = \lim_{h \to 0} \frac{f(a+h) - f(a)}{h}$$

if that limit exists. Equivalently, $f'(a) = (f_1'(a), f_2'(a), f_3'(a))$, if the component derivatives exist. We say that $f(t)$ is differentiable at a if $f'(a)$ exists. Recall that the derivative of a real-valued function of a single variable is a real number, representing the slope of the tangent line to the graph of the function at a point. Similarly, the derivative of a vector-valued function is a tangent vector to the curve in space which the function represents, and it lies on the *tangent line* to the curve.

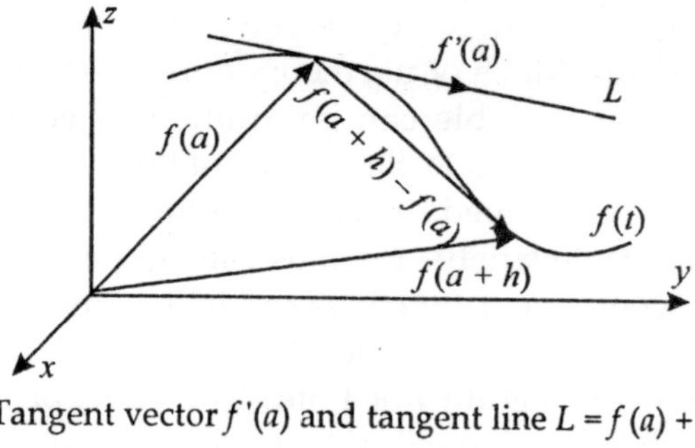

Fig. Tangent vector $f'(a)$ and tangent line $L = f(a) + sf'(a)$

Example: Let $f(t) = (\cos t, \sin t, t)$. Then $f'(t) = (-\sin t, \cos t, 1)$ for all t. The tangent line L to the curve at $f(2\pi) = (1, 0, 2\pi)$ is $L = f'(2\pi) + s f'(2\pi) = (1, 0, 2\pi) + s(0, 1, 1)$, or in parametric form: $x = 1, y = s, z = 2\pi + s$ for $-\infty < s < \infty$. A scalar function is a real-valued function. Note that if $u(t)$ is a scalar function and $f(t)$ is a vector-valued function, then their product, defined by $(u f)(t) = u(t) f(t)$ for all t, is a vector-valued function (since the product of a scalar with a vector is a vector).

3

Calculus of Vectors

In this chapter we learn about Vector Calculus *i.e.* Differentiation and Integration of vectors and their application.

LIMIT AND CONTINUITY

Let a vector function f be defined for all values of t in some neigthbourhood about a point t_0 possibly at t_0. Then, a vector a is called the limit of f as $t \to t_0$ if given any $\varepsilon > 0$, however small, there exists a real number $\delta > 0$ such that,

$$|f(t) - a| < \varepsilon, \text{ whenever } 0 < |t - t_0| < \delta$$

and we write,

$$\lim_{t \to t_0} f(t) = a \text{ or } \underset{t \to t_0}{Lt} f(t) = a.$$

The same definition holds for the limit of a scalar function if f is replaced by a scalar function f, and vector a by a scalar a. We state below some important theorems on limits of a vector function.

Theorem: If $f = f_1 i + f_2 j + f_3 k$ and $a = a_1 i + a_2 j + a_3 k$, then

$$\lim_{t \to t_0} f(t) = a$$

if and only if,

$$\lim_{t \to t_0} f_i(t) = a_i, i = 1, 2, 3.$$

Theorem: Let f and g be vector functions of a scalar variable t. Let f be a scalar function of t.

Then,

1. $$\underset{t \to t_0}{Lt} [f(t) + g(t)] = \underset{t \to t_0}{Lt} f(t) + \underset{t \to t_0}{Lt} g(t)$$

2. $\underset{t \to t_0}{Lt}[f(t) - g(t)] = \underset{t \to t_0}{Lt} f(t) - \underset{t \to t_0}{Lt} g(t)$

3. $\underset{t \to t_0}{Lt}[f(t) \cdot g(t)] = \left\{\underset{t \to t_0}{Lt} f(t)\right\} \cdot \left\{\underset{t \to t_0}{Lt} g(t)\right\}$

4. $\underset{t \to t_0}{Lt}[f(t) \times g(t)] = \left\{\underset{t \to t_0}{Lt} f(t)\right\} \times \left\{\underset{t \to t_0}{Lt} g(t)\right\}$

5. $\underset{t \to t_0}{Lt}[\phi(t) \times g(t)] = \left\{\underset{t \to t_0}{Lt} \phi(t)\right\} \times \left\{\underset{t \to t_0}{Lt} g(t)\right\}$

A vector function f is said to be continuous at $t = t_0$ if it is defined in some neigthbourhood of t_0 and,

$$\underset{t \to t_0}{Lt} f(t) = \underset{t \to t_0}{Lt} f(t)$$

Theorem: If $f = f_1 i + f_2 j + f_3 k$ is a function of t, then f is continuous at t_0 if and only if f_1, f_2 and f_3 are all continuous at t_0.

DIÛERENTIATION OF VECTOR-VALUED FUNCTIONS

A curve C is deûned by $r = r(t)$, a vector-valued function of one (scalar) variable. Let us imagine that C is the path taken by a particle and t is time. The vector $r(t)$ is the position vector of the particle at time t and $r(t + h)$ is the position vector at a later time $t + h$. The average velocity of the particle in the time interval $[t, t + h]$ is then,

$$\frac{\text{displacement vector}}{\text{lengh of time interval}} = \frac{r(t+h) - r(t)}{h}$$

In terms of the components of r this is,

$$\left(\frac{x(t+h) - x(t)}{h}, \frac{y(t+h) - y(t)}{h}, \frac{z(t+h) - z(t)}{h}\right)$$

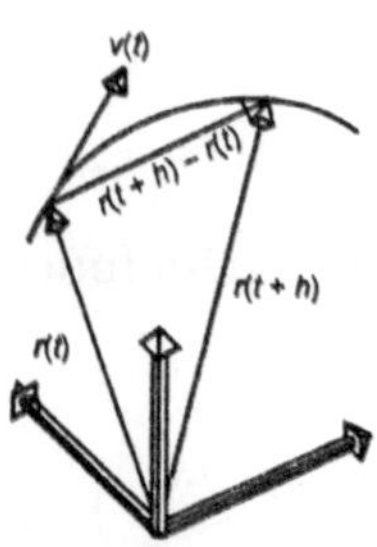

If each of the scalar function x, y and z are diûerentiable, then this vector has a limit,

$$\left(\frac{dx}{dt}, \frac{dy}{dt}, \frac{dz}{dt}\right) = (\dot{x}, \dot{y}, \dot{z})$$

which is the instantaneous velocity of the particle $v(t)$. This means that (if the motion is smooth) then,

$$v(t) = \lim_{h \to} \frac{r(t+h) - r(t)}{h} = \frac{d}{dt} r(t)$$

Theorem: Let $f(t)$ and $g(t)$ be differentiable vector-valued functions, let $u(t)$ be a differentiable scalar function, let k be a scalar, and let c be a constant vector. Then,

(a) $\frac{d}{dt}(c) = 0$

(b) $\frac{d}{dt}(kf) = k\frac{df}{dt}$

(c) $\frac{d}{dt}(f+g) = \frac{df}{dt} + \frac{dg}{dt}$

(d) $\frac{d}{dt}(f-g) = \frac{df}{dt} - \frac{dg}{dt}$

(e) $\frac{d}{dt}(uf) = \frac{du}{dt} f + u\frac{df}{dt}$

(f) $\frac{d}{dt}(f.g) = \frac{df}{dt}.g + f.\frac{dg}{dt}$

(g) $\frac{d}{dt}(f \times g) = \frac{df}{dt} \times g + f \times \frac{dg}{dt}$

Proof: The proofs of parts (a)-(e) follow easily by differentiating the component functions and using the rules for derivatives from single-variable calculus. We will prove part *(f)*, and leave the proof of part *(g)* as an exercise for the reader.

(f) Write $f(t) = (f_1(t), f_2(t), f_3(t))$ and $g(t) = (g_1(t), g_2(t), g_3(t))$, where the component functions $f_1(t)$, $f_2(t)$, $f_3(t)$, $g_1(t)$, $g_2(t)$, $g_3(t)$ are all differentiable real-valued functions. Then,

$$\frac{d}{dt}(f(t).g(t))$$

$$= \frac{d}{dt}(f_1(t)\, g_1(t) + f_2(t)\, g_2(t) + f_3(t)\, g_3(t))$$

$$= \frac{d}{dt}(f_1(t)g_1(t)) + \frac{d}{dt}(f_2(t)g_2(t) + \frac{d}{dt}(f_3(t)g_3(t))$$

$$= \frac{df_1}{dt}(t)g_1(t) + f_1(t)\frac{dg_1}{dt}(t) + \frac{df_2}{dt}(t)g_2(t) + f_2(t)$$

$$\frac{dg_2}{dt} + \frac{df_3}{dt}(t)g_3(t) + f_3(t)\frac{dg_3}{dt}(t)$$

$$= \left(\frac{df_1}{dt}(t), \frac{df_1}{dt}(t), \frac{df_1}{dt}(t)\right).(g_1(t), g_2(t), g_3(t))$$

$$+(f_1(t), f_2(t), f_3(t)).\left(\frac{dg_1}{dt}(t), \frac{dg_1}{dt}(t), \frac{dg_1}{dt}(t)\right)$$

$$= \frac{df}{dt}(t).g(t) + f(t).\frac{dg}{dt}(t) \text{ for all } t.$$

RULES FOR DIFFERENTIATION

To differentiate the vector product of vector function, the order of factors must be preserved because the vector product is not commutative.. If f, g and h are differentiable vector functions and ϕ is a scalar function of a scalar variable t, then

1. $\frac{d}{dt}(f \pm g) = \frac{df}{dt} \pm \frac{dg}{dt}$
2. $\frac{d}{dt}(f \cdot g) = f \cdot \frac{dg}{dt} + g \cdot \frac{df}{dt}$ [Since $f.g = g.f$, \in evaluating $\frac{d}{dt}(f \cdot g)$, we do not care about the order of f and g]
3. $\frac{d}{dt}(f \times g) = f \times \frac{dg}{dt} + \frac{df}{dt} \times g$ [Since $f \times g = -g'f$, $\therefore$ in evaluating $\frac{d}{dt}(f \times g)$, we do not care about the order of f and g]
4. $\frac{d}{dt}(\phi f) = \phi \frac{df}{dt} + f \frac{d\phi}{dt}$
5. $\frac{d}{dt}[fgh] = \left[\frac{df}{dt}gh\right] + \left[f\frac{dg}{dt}h\right] + \left[fg\frac{dh}{dt}\right]$

   Where [fgh] denotes the scalar triple product of f, g and h. [In evaluating,

$\frac{d}{dt}[fgh]$, we must keep the cyclic order of *f*, *g* and h in tact]

6. $\frac{d}{dt}(f\times(g\times h))=\frac{df}{dt}\times(g\times h)+f\times\left[\frac{dg}{dt}\times h\right]+f\times\left[g\times\frac{dh}{dt}\right]$

Derivative of a Function of a Function (Chain rule)

Let *r* be a differentiable function of a scalar variables *s* and *s* be a differentiable function of another scalar variable *t*, then *r* is a differentiable function of *t*. Let δt be a small change in *t* and δs, δr denote the corresponding changes in *s* and *r* respectively.

Now,

$$\frac{\delta r}{\delta t}=\frac{\delta r}{\delta s}\cdot\frac{\delta s}{\delta t}$$

Taking limits as $\delta t \to 0$ (and hence $\delta s \to 0$), we have,

$$\underset{\delta t\to 0}{Lt}\frac{\delta r}{\delta t}=\left(\underset{\delta t\to 0}{Lt}\frac{\delta r}{\delta s}\right)\cdot\left(\underset{\delta t\to 0}{Lt}\frac{\delta s}{\delta t}\right)\text{ or }\frac{dr}{dt}=\frac{dr}{ds}\cdot\frac{ds}{dt}$$

Constant Vectors

A vector is said to be constant if both its magnitude and direction do not change.

Partial Derivatives of Vector Functions of more than one Variable

So far, we have considered vector functions of a single scalar variable t. But, we come across with many vector functions of more than one independent scalar variable. If f is a vector function of independent variables x, y and z, we write $f=f(x, y, z)$. Let δx be a small change in *x*, while other variables remain unchanged. Let *f* be the corresponding change in *f*,

Then, $\delta f=f(x+\delta x, y, z)-f(x, y, z)$

Dividing by δx both sides we get,

$$\frac{\delta f}{\delta x}=\frac{f(x+\delta x,y,z)-f(x,y,z)}{\delta x}$$

Taking limits $\delta x \to 0$, we have,

$$\underset{\delta x\to 0}{Lt}\frac{\delta f}{\delta x}=\underset{\delta x\to 0}{Lt}\frac{f(x+\delta x,y,z)-f(x,y,z)}{\delta x}$$

If this limit exists, we denote it by,

$$\frac{\partial f}{\partial x}$$

or f_x and is called the partial derivative of f *w. r. t* x *i.e.,*

$$\frac{\partial f}{\partial x} = \underset{\delta x \to 0}{Lt} \frac{\delta f}{\delta x} = \underset{\delta x \to 0}{Lt} \frac{f(x+\delta x, y, z) - f(x,y,z)}{\delta x}$$

Similarly we define the partial derivatives,

$$\frac{\partial f}{\partial y} = \underset{\delta y \to 0}{Lt} \frac{\delta f}{\delta y} = \underset{\delta y \to 0}{Lt} \frac{f(x, y+\delta y, z) - f(x,y,z)}{\delta y}$$

and,

$$\frac{\partial f}{\partial z} = \underset{\delta z \to 0}{Lt} \frac{\delta f}{\delta z} = \underset{\delta z \to 0}{Lt} \frac{f(x, y, z+\delta z) - f(x,y,z)}{\delta z}$$

Higher –order partial derivatives may also be defined in the same way. For example,

$$f_{xx} = \frac{\partial^2 f}{\partial x^2} = \frac{\partial}{\partial x}\left(\frac{\partial f}{\partial x}\right) \quad f_{yy} = \frac{\partial^2 f}{\partial y^2} = \frac{\partial}{\partial y}\left(\frac{\partial f}{\partial y}\right)$$

$$f_{xy} = \frac{\partial}{\partial y}\left(\frac{\partial f}{\partial x}\right) = \frac{\partial^2 f}{\partial y \partial x}, \quad f_{yx} = \frac{\partial}{\partial x}\left(\frac{\partial f}{\partial y}\right) = \frac{\partial^2 f}{\partial x \partial y}$$

If f has continuous partial derivatives of second order, then,

$$\frac{\partial^2 f}{\partial x \partial y} = \frac{\partial^2 f}{\partial y \partial x}.$$

Rules for Partial Differentiation

Rules for partial differentiation of vectors are similar to those used in elementary calculus for scalar functions. Thus, if f and g are differentiable vector functions of x, y, z and ϕ is a differentiable scalar function of x, y, z then (using x as an example)

1. $\frac{\partial}{\partial x}(f \pm g) = \frac{\partial f}{\partial x} \pm \frac{\partial g}{\partial x}$
2. $\frac{\partial}{\partial x}(f \cdot g) = f \cdot \frac{\partial g}{\partial x} + g \cdot \frac{\partial f}{\partial x}$
3. $\frac{\partial}{\partial x}(f \times g) = f \times \frac{\partial g}{\partial x} + \frac{\partial f}{\partial x} \times g$

4. $\frac{\partial}{\partial x}(\phi f) = \phi \frac{\partial f}{\partial x} + f \frac{\partial \phi}{\partial x}$

5. $\frac{\partial^2}{\partial y \partial x}(f \cdot g) = \frac{\partial}{\partial y}\left\{\frac{\partial}{\partial x}(f \cdot g)\right\} = \frac{\partial}{\partial y}\left\{f \cdot \frac{\partial g}{\partial x} + g \cdot \frac{\partial f}{\partial x}\right\}$

$= \frac{\partial}{\partial y}\left(f \cdot \frac{\partial g}{\partial x}\right) + \frac{\partial}{\partial y}\left(g \cdot \frac{\partial f}{\partial x}\right)$

$= f \cdot \frac{\partial^2 g}{\partial y \partial x} + \frac{\partial f}{\partial x} \cdot \frac{\partial g}{\partial x} + \frac{\partial f}{\partial x} \cdot \frac{\partial g}{\partial x} + \frac{\partial^2 f}{\partial y \partial x} \cdot g$

6. $\frac{\partial^2}{\partial x \partial y}(f \times g) = \frac{\partial}{\partial x}\left\{\frac{\partial}{\partial y}(f \times g)\right\} = \frac{\partial}{\partial x}\left\{f \times \frac{\partial g}{\partial y} + \frac{\partial f}{\partial y} \times g\right\}$

$= \frac{\partial}{\partial x}\left(f \times \frac{\partial g}{\partial y}\right) + \frac{\partial}{\partial x}\left(\frac{\partial f}{\partial y} \times g\right)$

$= f \times \frac{\partial^2 g}{\partial x \partial y} + \frac{\partial f}{\partial x} \times \frac{\partial g}{\partial y} + \frac{\partial f}{\partial y} \times \frac{\partial g}{\partial x} + \frac{\partial^2 f}{\partial x \partial y} \times g$

If $f(x,y,z) = f_1(x,y,z)i + f_2(x,y,z)j + f_3(x,y,z)k$, then,

$$\frac{\partial f}{\partial x} = \frac{\partial f_1}{\partial x} i + \frac{\partial f_2}{\partial x} j + \frac{\partial f_3}{\partial x} k \text{ etc}$$

If a vector function f is a function of *x*, *y*, *z* and the variables *x*, *y*, *z* are expressed in terms of new variables *s*, *t*, then the function will be a function of *s* and *t*. Then, we can write,

$$\frac{\partial f}{\partial t} = \frac{\partial f}{\partial x} \cdot \frac{\partial x}{\partial t} + \frac{\partial f}{\partial y} \cdot \frac{\partial y}{\partial t} + \frac{\partial f}{\partial z} \cdot \frac{\partial z}{\partial t} \text{ etc.}$$

Total Differentials

Differentials of vectors follow rules similar to those of elementary calculus.

For example,

1. If,
   $f(x,y,z) = f_1(x,y,z)i + f_2(x,y,z)j + f_3(x,y,z)k$,
   then,
   $df(x,y,z) = df_1(x,y,z)i + df_2(x,y,z)j + df_3(x,y,z)k$
2. $d(f \cdot g) = f \cdot dg + g \cdot df$
3. $d(f \times g) = f \times dg + df \times g$
4. If $f = f(x, y, z)$, then df,

$$= \frac{\partial f}{\partial x}dx + \frac{\partial f}{\partial y}dy + \frac{\partial f}{\partial z}dz \text{, etc}$$

Space Curve

If $r = f(t)$ denotes a vector function of a scalar variable t and r is the position vector of a point $P(x, y, z)$, in space, then as t varies, so does r and therefore the end point P or r describes a continuous curve. Therefore, a curve C can always be represented by expressing the position vector r of any point on it as a function of some scalar variable t.

If a $\leq t \leq b$, the end point of the vector $r(a)$ is called initial point of the curve C and the end point of the vector $r(b)$ is called the end point of C. If $r(a) = r(b)$, then curve C is a closed curve. That is, a curve is closed if its initial and end point coincide. If there exist $a < t_1$ and $t_2 < b$ such that $r(t_1) = r(t_2) = r_0$ (say), then r is called a multiple point. Geometrically, a multiple point occurs where curve C crosses itself or is tangent to itself. A curve C is called a simple curve if it has no multiple points. A curve which is both simple and closed, is called a simple closed curve. Let P be a fixed point on the curve having position vector r corresponding to some t. Let Q be the point corresponding to $t + \delta t$ and have position vector $r + \delta r$.

Then,

$$\underset{\delta t \to 0}{Lt} \frac{PQ}{\delta t} = \underset{\delta t \to 0}{Lt} \frac{\delta r}{\delta s} = \frac{dr}{dt}$$

is a vector tangent to the space curve C at P. This vector is called the tangent vector to C at P. A point t on a space curve C is called singular or non-singular according as the value of

$$\frac{dr}{dt}$$

is zero or non-zero at t. The direction of the space curve C at a non-singular point P is the same as that of the tangent vector to C at P. A vector function which has a continuous derivative and no singular points is called a smooth vector function. If the parameter t is the time, then represents,

$$\frac{dr}{dt}$$

the velocity v, with which the end point of r describes the curve and,

$$\frac{dv}{dt} = \frac{d^2r}{dt^2}$$

represents its acceleration a along the curve.

INTEGRATION OF VECTORS

We have learnt about differentiation of vector valued function and their properties with their application. In this chapter we study the integration of vector function and its application.

Integration of Vector Function

If f and F are two vector functions of the scalar variable t such that,

$$\frac{dF}{dt} = f,$$

Then F is called the indefinite integral of $w.\ r.\ t.$ and we write. The function f, which is to be integrated is called the integrand. If c is an arbitrary constant vector independent of t, then,

$$\frac{d}{dt}(F + c) = \frac{dF}{dt} = f$$

$$\therefore \int f dt = F + c$$

Thus, the integral F is indefinite to the extent of an additive arbitrary constant vector c.

If,

$$\frac{dF}{dt} = f$$

for all values of t in the interval $[a, b]$, the definite integral between the limits $t = a$ and $t = b$ is written as

$$\int_a^b f dt = [F]_a^b = F(b) - F(a)$$

Theorem: If $f = f_1\, i + f_2\, j + f_3\, k$, where, f_1, f_2, f_3 are scalars functions of the scalar variable t, then

$$\int f dt = i\int f_1 dt + j\int f_2 dt + k\int f_3 dt$$

Thus, to integrate a vector function, we should integrate its components.

Some Standard Results

We are going to dicuss some standard results for integration:

Let f and g be two vector function of scalar variable t.

1. Since,

$$\frac{d}{dt}(f \cdot g) = \frac{df}{dt} \cdot g + f \cdot \frac{df}{dt}$$

integrating both side, we get,

$$\int\left(\frac{df}{dt}.g + f.\frac{df}{dt}\right)dt = f.g + c$$

Where c is the constant of integration and is a scalar because the integrand is a scalar.

2. Since,

$$\frac{d}{dt}(f \cdot f) = \frac{d}{dt}(f \cdot f) = \frac{df}{dt} \cdot f + f \cdot \frac{d}{dt} = 2f \cdot \frac{df}{dt}$$

$$\therefore \int 2f \cdot \frac{df}{dt} = f^2 + c$$

Where c is the constant of integration and is a scalar because the integrand is a scalar.

3. Since,

$$\frac{d}{dt}\left(\frac{df}{dt}\right)^2 = 2\frac{df}{dt} \cdot \frac{d^2 f}{dt^2}$$

$$\therefore \quad \int\left(2\frac{df}{dt} \cdot \frac{d^2 f}{dt^2} g\right)dt = \left(\frac{df}{dt}\right) + c$$

Where c is the constant of integration and is a scalar because the integrand is a scalar.

4. Since,

$$\frac{d}{dt}(f \times g) = \frac{df}{dt} \times g + f \times \frac{dg}{dt}$$

$$\therefore \quad \int\left(\frac{df}{dt} \times g + f \times \frac{dg}{dt}\right)dt = f \times g + c$$

Where c is the constant of integration and is a scalar because the integrand is a scalar

5. Since,

$$\frac{d}{dt}(a \times g) = a \times \frac{df}{dt},$$

where a is constant vector.

$$\therefore \qquad \int \left(a \times \frac{df}{dt} \right) dt = a \times f + c$$

Where c is the constant of integration.

6. If c is a constant scalar,

$$\int cf dt = c \int f dt$$

7. $\int (f \pm g) dt = \int f dt \pm \int g dt$

Remember: The constant of integration is of the same nature as the integrand

Example: If $r = (7t^2)\, i + 2\, t\, j - 3\, t^2\, k$, find,

$$\int_2^3 r \times \frac{d^2 r}{dt^2} dt.$$

Solution: Since,

$$\frac{d}{dt}\left[r \times \frac{dr}{dt} \right] = r \times \frac{d^2 r}{dt^2} + \frac{dr}{dt} \times \frac{dr}{dt}$$

$$\Rightarrow \qquad \frac{d}{dt}\left[r \times \frac{dr}{dt} \right] = r \times \frac{d^2 r}{dt^2}$$

$\therefore$ Integrating both sides *w. r. t.*, we have

$$\int \left(r \times \frac{d^2 r}{dt^2} \right) dt = r \times \frac{dr}{dt} + c$$

where c is the constant of integration.

$$\therefore \qquad \int_2^3 \left(r \times \frac{d^2 r}{dt^2} \right) dt = \left[r \times \frac{dr}{dt} \right]_2^3$$

Let us find $r \times \frac{dr}{dt}$

Since, $r = (7t^2)\, i + 2\, t\, j - 3\, t^2\, k$

$$\therefore \qquad \frac{dr}{dt} = \frac{d}{dt} \{(7t^2)\, i + 2\, t\, j + (-3\, t^2)\, k\}$$

$$\therefore \qquad \frac{dr}{dt} = 14\, t\, I + 2\, j - 6\, t\, k$$

$$\therefore \quad \left[r\times\frac{dr}{dt}\right] = \begin{vmatrix} i & j & k \\ 7t^2 & 2t & -3t^2 \\ 14t & 2 & -6t \end{vmatrix} = i\left(-12t^2+6t^2\right)$$

$$+j\left(-42t^3+42t^3\right)+k\left(14t^2-28t^2\right)$$

$$\Rightarrow \quad \left[r\times\frac{dr}{dt}\right] = i\left(-6t^2\right)+j(0)+k\left(-14t^2\right)$$

Using (2) in (1), we get

$$\therefore \quad \int_2^3\left(r\times\frac{d^2r}{dt^2}\right)dt = \left[i\left(-6t^2\right)+j(0)+k\left(-14t^2\right)\right]_2^3$$

$$= \left[i\left\{\left(-6.3^2\right)-\left(-6.2^2\right)\right\}+k\left\{\left(-14.3^2\right)-\left(-14.2^2\right)\right\}\right]$$

Hence, $\int_2^3 r\times\frac{d^2r}{dt^2}dt = -30\,i-70\,k$

4

Integration of Vectors Function

In the previous chapter we have learnt about differentiation of vector valued function and their properties with their application. In this chapter we study the integration of vector function and its application..

Integration of Vector Function

If f and F are two vector functions of the scalar variable t such that,

$$\frac{dF}{dt} = f,$$

Then F is called the indefinite integral of *w. r .t.* and we write, the function f, which is to be integrated is called the integrand. If c is an arbitrary constant vector independent of t, then,

$$\frac{d}{dt}(F + c) = \frac{dF}{dt} = f \therefore \int f dt = F + c$$

Thus, the integral *F* is indefinite to the extent of an additive arbitrary constant vector *c*.

If,

$$\frac{dF}{dt} = f$$

for all values of t in the interval [*a, b*], the definite integral between the limits $t = a$ and $t = b$ is written as,

$$\int_a^b f dt = [F]_a^b = F(b) - F(a)$$

Theorem. If $f = f_1\, i + f_2\, j + f_3\, k$, where, f_1, f_2, f_3 are scalars functions of the scalar variable *t*, then,

$$\int f dt = i\int f_1 dt + j\int f_2 dt + k\int f_3 dt$$

Thus, to integrate a vector function, we should integrate its components.

CURVE IN A VECTOR FIELD

Let I be an interval of real numbers (*i.e.* a non-empty connected subset of R). Then a curve g is a continuous mapping $\gamma : I \to X$, where X is a real space.

The curve g is said to be simple if for all x, y in I, we have $\gamma(x) = \gamma(y) \to x = y$. If I is a closed bounded interval $[a, b]$ and $\gamma(a) = \gamma(b)$ then the curve is closed simple curve). If $\gamma(x) = \gamma(y)$ for some $x \neq y$ (other than the extremities of I), then $\gamma(x)$ is called a double (or multiple) point of the curve.

A curve γ is said to be closed or a loop if $I = (a, b)$ and if $\gamma(a) = \gamma(b)$. A closed curve is thus a continuous mapping of the circle S_1; a simple closed curve is also called a Jordan curve.

LINE INTEGRALS

Suppose that a wire has as density $f(x, y, z)$ at the point (x, y, z) on the wire. Then the line integral will equal the total mass of the wire. Below is the definition in symbols.

Definition of the Line Integral

Let f be a function defined on a curve C of finite length. Then the *line integral* of f along C is

$$\int_C f(x,y)ds = \lim_{n\to\infty} \sum_{i=1}^{n} f(x_i, y_i)\Delta s_i \quad \text{(for two dimension)}$$

$$\int_C f(x,y,z)ds = \lim_{n\to\infty} \sum_{i=1}^{n} f(x_i, y_i, z_i)\Delta s_i \quad \text{(for three dimension)}$$

Evaluating Line Integrals

Let, $r(t) = x(t)i + y(t)j$

be a differentiable vector valued function. Then

$$ds = \|r'(t)\|dt = \sqrt{x'(t)^2 + y'(t)^2}\,dt$$

We are now ready to state the theorem that shows us how to compute a line integral.

Theorem: Line Integrals Over Vector Valued Functions

Let $r(t) = x(t)i + y(t)j \quad a \leq t \leq b$

be a differentiable vector valued function that defines a smooth curve C.

Then,

$$\int_C f(x,y)ds = \int_a^b f(x(t),y(t))\sqrt{x'(t)^2+y'(t)^2}dt$$

and for three dimensions, if,

$$r(t) = x(t)i + y(t)j + z(t)k \quad a \le t \le b$$

then,

$$\int_C f(x,y,z)ds$$

$$= \int_a^b f(x(t),y(t),z(t))\sqrt{x'(t)^2+y'(t)^2+z'(t)^2}dt$$

Line Integrals in Differential Form

We can rewrite $r'(t)$ dt as

$$\frac{dr}{dt}dt = \left(\frac{dx}{dt}i + \frac{dy}{dt}j + \frac{dz}{dt}k\right)dt$$

$$\Rightarrow dr = dxi + dyj + dzk$$

So that if

$$F = Mi + Nj + Pk \text{ then}$$

$$F \cdot r'(t)dt = Mdx + Ndy + Pdz$$

This is called the differential form. Let $r = xi + yj + zk$ be the position vector of appoint $P(x, y, z)$ on a curve C, where x, y, z are functions of scalar parameter t, $a \le t \le b$.

Let $t = a$ at A and $t = b$ at B. Let $f = f_1 i + f_2 j + f_3 k$ be a vector point function defined and continuous along curve C. If s denotes the arc length of the curve C, then $\frac{dr}{dt} = t$ is a unit vector along the tangent to the curve C at P. The component of the vector f along the tangent at P is

$$f_t = f \cdot t = f \cdot \frac{dr}{ds}$$

For points on C, f_t is a unction of s and $\int_A^B f_t ds$ is known as the tangent line integral or just line integral of f along C.

But $f_t = f \cdot t = f \cdot \frac{dr}{ds}$ ds = f.dr

$$\therefore \int_A^B f_t ds = \int_A^B f \cdot t ds = \int_A^B f \cdot dr$$

and is also written as $\int_C f \cdot dr$

Note that $\int_A^B f \cdot dr = -\int_B^A f \cdot dr$

The differential displacement vector dr along the curve C is defined by,

$dr = dxi + dyj + dzk$

$f \cdot dr = (f_1 i + f_2 j + f_3 k) \cdot (dxi + dyj + dzk)$

$= f_1 dx + f_2 dy + f_3 dz$

Therefore , in component form the above line can be written as,

$$\int_C f \cdot dr = \int_C (f_1 i + f_2 j + f_3 k) \cdot (dxi + dyj + dzk)$$

$$= \int_C (f_1 dx + f_2 dy + f_3 dz)$$

If the given curve is given by the parametric equations: $x = x(t), y = y(t), z = z(t)$, then can be write

$$\int_C f \cdot dr = \int_a^b \left(f_1 \frac{dx}{dt} + f_2 \frac{dy}{dt} + f_3 \frac{dz}{dt} \right) dt,$$

Where *a*, *b* are the values of the parameter *t* at *A* and *B*.

Circulation

If C is a simple closed curve, then the tangent line integral of f around C is called the circulation of about C and is often denoted by

$$\oint_C f \cdot dr = \oint_C (f_1 dx + f_2 dy + f_3 dz)$$

Applications of line integral

The main application of line integrals is finding the work done on an object in a force field. The work done W along each piece will be approximately equal to

$dW = F \cdot T\, ds$ Now recall that,

$$T = \frac{r'(t)}{\|r'(t)\|}$$

and that

$$ds = ||r'(t)||dt$$

Hence,

$$dW = F \cdot r'(t)dt$$

As usual, we add up all the small pieces of work and take the limit as the pieces get small to end up with an integral.

Definition: Let F be a vector field and C be a curve defined by the vector valued function r. Then the work done by F on an object moving along C is given by Work done =

$$\int_C F \cdot dr = \int_a^b F(x(t), y(t), z(t)) \cdot r'(t)dt$$

SURFACE INTEGRAL

Let S be a surface in three-dimensional space R^3 and let f be a function defined on this surface. Then the *integral of f with respect to the surface area* is usually notated as

$$\int_S f(u,v)d^2A$$

To compute above integral, one may use the following formula to convert it to a double integral:

$$\int_S f(u,v)d^2A$$

$$= \int f(u,v)\sqrt{\left(\frac{\partial(x,y)}{\partial(u,v)}\right)^2 + \left(\frac{\partial(y,z)}{\partial(u,v)}\right)^2 + \left(\frac{\partial(z,x)}{\partial(u,v)}\right)^2}\, dudv$$

If the surface is described as the graph of a function g, then we may also use the following formula:

$$\int_S f(x,y)d^2A = \int f(x,y)\sqrt{1 + \left(\frac{\partial g}{\partial x}\right)^2 + \left(\frac{\partial g}{\partial y}\right)^2}\, dxdy$$

Evaluating Surface Integral

For a scalar function f over a surface parameterized by u and v, the surface integral is given by

$$\phi = \int_S f d\alpha = \int_S f(u,v)|T_u \times T_v|\, dudv$$

where T_u and T_v are tangent vectors and a ′ b is the cross product. Let P be any point on the surface S and let n be the unit vector at P in the direction of the outward drawn normal to the surface S at P. Then, $f \cdot n$ is the normal component of f at P. The integral of $f \cdot n$ over S is called the (normal) surface integral of f over S and is denoted by

$$\iint_S f \cdot ndS$$

It is also called the flux of f over S. If we associate with the differential of surface area dS a vector dS whose magnitude is dS and whose direction is that of n, then, $dS = ndS$ where $f_n = f \cdot n$ denotes the component of f in the direction of the outward normal at P. If the outward drawn normal to the surface S at P makes angles 2, 3, 4 with the positive direction of x, y, z axes respectively, and l, m, n are the direction cosines of this outward normal, then,

$$l = \cos\alpha, m = \cos\beta, n = \cos\gamma$$

Also, $\hat{n} = \cos\alpha\, i + \cos\beta\, j + \cos\gamma\, k = li + mj + nk$

If $f = f_1 i + f_2 j + f_3 k$, then $f \cdot \hat{n}$

$= (f_1 i + f_2 j + f_3 k) \cdot (li + mj + nk)$

$= f_1 l + f_2 m + f_3 n$

$= f_1 \cos\alpha + f_2 \cos\beta + f_3 \cos\gamma$

$$\iint_S f \cdot \hat{n}\, dS = \iint_S (f_1 \cos\alpha + f_2 \cos\beta + f_3 \cos\gamma) dS$$

Now, $dS \cos\alpha$ = orthogonal projection of the area of the dS on the yz–plane = $dy\, dz$ $dS \cos\beta$ = orthogonal projection of the area of the dS on the yz–plane = $dz\, dx$: and $dS \cos\gamma$ = orthogonal projection of the area of the dS on the yz –plane = $dx\, dy$

$$\therefore \iint_S f \cdot \hat{n}\, dS = \iint_S (f_1 dydz + f_2 dzdx + f_3 dxdy)$$

Also, we can evaluate the surface integral using the method given below.

$$\phi = \int_S F \cdot da = \int_S (F \cdot \hat{n})\, d\alpha$$

$$= \int_S \left(f_x dy\, dz + f_y dz\, dx + f_z dx\, dy\right)$$

where $a \cdot b$ is a dot product and $\hat{n}$ is a unit normal vector.

If $z = f(x, y)$, then $d\, a$ is given explicitly by,

$$da = \pm\left(-\frac{\partial z}{\partial x}\hat{x} - \frac{\partial z}{\partial y}\hat{y} + \hat{z}\right) dxdy$$

If the surface is surface parameterized using u and v, then

$$\phi = \int_S F\cdot(T_u \times T_v)\,dudv$$

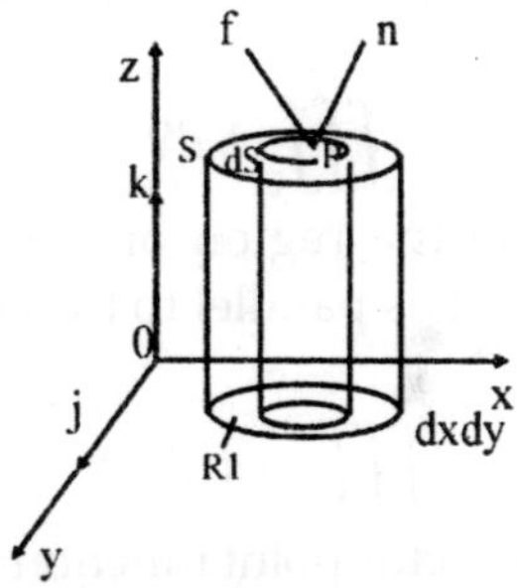

Let S be a surface such that any line perpendicular to xy–plane meets S in not more than one point. Then, the equation of the surface S can be written as $z = h(x, y)$. Let the projection of S on xy– plane = $dS\cos g$, where g is the acute angle which the normal n at P to the surface S makes with z- axis.

$$\therefore\ dx = \cos\gamma = dxdy$$

Also, $\cos\gamma = |n\cdot k|$, where k is unit vector along z-axis

$$\therefore\ dS = \frac{dxdy}{|n\cdot k|}$$

Hence,

$$\iint_S f\cdot n dS = \iint_{R_1} f\cdot n\frac{dxdy}{|n\cdot k|}.$$

Similarly,

$$\iint_S f\cdot ndS = \iint_{R_2} f\cdot n\frac{dydz}{|n\cdot i|} = \iint_{R_2} f\cdot n\frac{dzdx}{|n\cdot j|}$$

Volume Integral

Let be a scalar point function defined throughout a given region of volume V. Divide the given region into n sub-regions of volumes $\delta V_1, \delta V_2, \ldots, \delta V_n$. Let $P(x_i, y_i, z_i)$ be any point inside or on the boundary of the sub-regions of volume dV_i. From the sum,

$$\sum_{i=1}^{n} \phi(P_i)\delta V_i$$

where $\phi(P_i) = \phi(x_i, y_i, z_i)$. If as $n \to \infty$ in such a way that each

dV_i tends to zero, the sum tends to a finite limit and is independent of the mode of sub-division, then this limit is called the volume integral of f over the volume V and is denoted by,

$$\iiint_V \phi\, dV$$

If we sub-divide the region of volume V into small a cuboids by drawing lines parallel to the coordinate axes, then $dV = dx\, dy\, dz$ and,

$$\iiint_V \phi\, dxdydz$$

Likewise if f is a vector point function defined in the given region of volume V, we define and denote the vector volume integral of f over V by,

$$\iiint_V f\, dV\,.$$

A triple integral over three coordinates giving the volume within some region G,

$$V = \iiint_G \phi\, dxdydz$$

Volume element

A volume element is the differential element dV whose volume integral over some range in a given coordinate system gives the volume of a solid,

$$V = \iiint_G \phi\, dxdydz$$

In $\hat{A}^n$, the volume of the infinitesimal n-hypercube bounded by $dx_1, dx_2, dx_3, \ldots, dx_n$ has volume given by the wedge product

$$dV = dx_1 \wedge dx_2 \wedge dx_3 \wedge \ldots \wedge dx_n$$

The use of the antisymmetric wedge product instead of the symmetric product $dx_1, dx_2, dx_3, \ldots, dx_n$ is a technical refinement often omitted in informal usage. Dropping the wedges, the volume element for curvilinear coordinates in $\Re^3$ is given by,

$$\begin{aligned} dV &= \left|(h_1 u_1 du_1)\cdot(h_2 u_2 du_2)\times(h_3 u_3 du_3)\right| \\ &= h_1 h_2 h_3\; du_1 du_2 du_3 \end{aligned}$$

$$= \left| \frac{\partial r}{\partial u_1} \cdot \frac{\partial r}{\partial u_2} \times \frac{\partial r}{\partial u_3} \right| du_1 du_2 du_3 = \begin{vmatrix} \frac{\partial x}{\partial u_1} & \frac{\partial x}{\partial u_2} & \frac{\partial x}{\partial u_3} \\ \frac{\partial y}{\partial u_1} & \frac{\partial y}{\partial u_2} & \frac{\partial y}{\partial u_3} \\ \frac{\partial z}{\partial u_1} & \frac{\partial z}{\partial u_2} & \frac{\partial z}{\partial u_3} \end{vmatrix} du_1 du_2 du_3$$

$$= \left| \frac{\partial(x,y,z)}{\partial(u_1,u_2,u_3)} \right| du_1 du_2 du_3$$

where the latter is the Jacobian and the h_i are scale factors.

Reduction of Volume to Surface Integrals

One of the most important theorems in vector analysis is known as the Divergence Theorem, which is also sometimes called Gauss' Theorem. This is essentially just an application of the fundamental theorem of calculus,

$$\int_a^b \frac{df(x)}{dx} dx = f(b) - f(a)$$

This enables us to express the integral of the quantity,

$$\frac{df}{dx}$$

along an interval in terms of the values of f itself at the endpoints of that interval. Now suppose we are given a scalar function $f(x, y, z)$ throughout a region V enclosed by a surface S, and we want to evaluate the integral of the quantity, over this entire region. This can be written as,

$$\int_V \frac{\partial f}{\partial x} dV = \iiint_{xyz} \frac{\partial f}{\partial x}\, dxdydz$$

where the three integrals are evaluated over suitable ranges to cover the entire region V. For any fixed values of $y = y_0$ and $z = z_0$ the function $f(x, y_0, z_0)$ is totally dependent on x, so we can evaluate the integral along any line parallel to the x axis through the region V for any particular y_0, z_0 using the fundamental theorem of calculus

$$\int_a^b \frac{df(x,y_0,z_0)}{dx} dx = f(b,y_0,z_0) - f(a,y_0,z_0)$$

where a, b are the x values at which the line intersects the surface S. Hence, we can cover all the contributions by integrating the value of the function $f(x, y, z)$ over the entire surface S. Consequently, we can write our original integral as,

$$\int_V \frac{\partial f}{\partial x} dV = \int_S f \cos(n \cdot i)\, dS$$

where cos (n, i) denotes the cosine of the angle between a unit vector normal to the surface and a basis vector in the positive x direction.

Naturally, since the coordinate axes are symmetrical, we could arrive at analogous results with x replaced by either y or z, and i replaced by j or k respectively.

Now suppose we are given a vector field $F(x, y, z)$ with the components $F_x(x, y, z)$, $F_y(x, y, z)$, and $F_z(x, y, z)$. Using the above results and simple additivity we have

$$\int_V \left(\frac{\partial F_x}{\partial x} + \frac{\partial F_y}{\partial y} + \frac{\partial F_z}{\partial z} \right) dV$$

$$= \int_S [F_x \cos(n \cdot i) + F_y \cos(n \cdot j) + F_z \cos(n.k)] dS$$

The integrand on the left side is $\nabla \times F$, *i.e.* the divergence of F. Also, notice that $\cos(n \cdot i)$, $\cos(n \cdot j)$, and $\cos(n \cdot k)$ are the components of the normal unit vector n, so the integrand on the right side is simply $F \times n$, *i.e.*, the dot product of F and the unit normal to the surface. Hence we can express the Divergence Theorem in its familiar form

$$\int_V \nabla \cdot F dV = \int_S F \cdot n\, dS$$

Gauss's Divergence Theorem

Statement: The surface integral of the normal component of a vector f taken over a closed surface is equal to the integral of the divergence of f taken over the volume enclosed by the surface.

Symbolically: If f is a continuously differentiable vector point function and S is closed surface enclosing the volume V with an outward pointing unit normal, n, the Divergence.

Theorem says:

$$\iint_S F \cdot n \, dS = \iiint_V div \text{ F dV}$$

Or,

$$\iint_S F \cdot n \, d\sigma = \iiint_V div \text{ F dV}$$

We take $F = Mi + Nj + Pk$

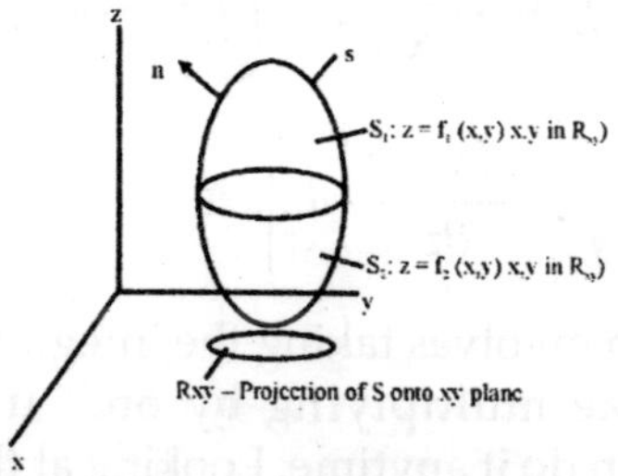

Using the above figure, we get

$n \cdot i = \cos A$–The angle from n to the x-axis

$n \cdot j = \cos B$–The angle from n to the y-axis

$n \cdot k = \cos C$–The angle from n to the z-axis

$n = \cos Ai + \cos Bj + \cos Ck$

since the dot product of n with each basis vector is that component of n. Now we restate the Divergence Theorem using the components of the vector field F.

$$\iint_S [M\cos A + N\cos B + P\cos C]\, d\sigma$$

$$= \iiint_V \left(\frac{\partial M}{\partial x} + \frac{\partial N}{\partial y} + \frac{\partial P}{\partial z}\right) dV$$

Taking just the z terms,

$$\iint_S P\cos C\, d\sigma = \iiint_V \frac{\partial P}{\partial z} dV$$

Splitting into integrals over the two surfaces S_1 and S_2,

$$\iint_{S_1} P(x,y,f_1(x,y))\cos C(x,y,f_1(x,y))\,dxdy$$

$$+\iint_{S_2} P(x,y,f_2(x,y))\cos C(x,y,f_2(x,y))\,dxdy$$

Project onto,

$$R_{xy}\ dS = \frac{dA}{\|n \cdot k\|} = \frac{dA}{|\cos C|}$$

On top(S_1) $\cos C \geq 0 \Rightarrow |\cos C| = \cos C$

On bottom (S_2) $\cos C \leq 0 \Rightarrow |\cos C| = -\cos C$

$$\iint_{R_{xy}} P(x,y,f_1(x,y))\cos C \frac{dA}{\cos C} - \iint_{S_2} P(x,y,f_2(x,y))\cos C \frac{dA}{\cos C}$$

$$= \iint_{R_{xy}} \left\{ \int_{f_2(x,y)}^{f_1(x,y)} \frac{\partial P(x,y,z)}{\partial z} dz \right\} dA$$

The last step involves taking the integral of the derivative of P. This is like multiplying by one. It doesn't change anything, you can do it anytime. Looking at the last term, what we really have is a volume integral. Evaluating the inner integral, we have $[f_1(x, y) - f_2(x, y)]\ '\ dA$. This is the volume of a differential element of the region. Adding all of them up gives us the volume of the region:

$$\iiint_V \frac{\partial P}{\partial V} dV$$

The other terms work out in a similar manner. If the region is not a convex region, we can construct it by placing convex regions next to each other and summing up the results. Thus, we have proved the Divergence Theorem for a large number of regions

Deduction from Gauss's theorem

1. $\iiint_V f \cdot \nabla\phi\ dV = \iint_S \phi f \cdot n dS - \iiint_V \phi\ \text{div} f\ dV$
2. $\iint_S f \times n\ dS = -\iiint_V curlf\ dV$
3. $\iint_S \phi\ n\ dS = \iiint_V grad\phi\ dV$

GREEN'S THEOREM

Let S be a sufficiently nice region in the plane, and let δS be its boundary; then we have,

$$\iint_S \left(\frac{\partial v_2}{\partial x} - \frac{\partial v_1}{\partial y}\right) dxdy = \oint_{\delta S} (v_1 dx + v_2 dy)$$

where the boundary, d*S* is traversed counterclockwise on its outside cycle, (and clockwise on any internal cycles).

Statement: If ϕ and ψ are two scalar point functions having continuous second order derivatives in a region *V* bounded by a closed surface *S*, then

$$\iiint_V \left(\phi\nabla^2\psi - \psi\nabla^2\phi\right) dv = \iint_S (\phi\nabla\psi - \psi\nabla\phi)\cdot n\, dS$$

Proof: By Gauss divergence theorem, we have,

$$\iiint_V \nabla \cdot f dv = \iint_S f \cdot n\, dS$$

Putting $f = \phi\nabla\psi$, we get,

$$\nabla \cdot f = \nabla \cdot (\phi\nabla\psi) = \phi(\nabla \cdot \nabla\psi) + (\nabla\phi)\cdot(\nabla\psi)$$
$$= \phi\left(\nabla^2\psi\right) + (\nabla\phi)\cdot(\nabla\psi)$$

Also, $f \cdot n = (\phi\nabla\psi)\cdot n$

$\therefore$ (68), becomes

$$\iiint_V \left[\phi\nabla^2\psi + (\nabla\phi)\cdot(\nabla\psi)\right] dv = \iint_S (\phi\nabla\psi)\cdot n\, dS$$

This is called the Green's first theorem (or identity) Interchanging ϕ and ψ in (69), we get,

$$\iiint_V \left[\psi\nabla^2\phi + (\nabla\psi)\cdot(\nabla\phi)\right] dv = \iint_S (\psi\nabla\phi)\cdot n\, dS$$

Operate (70)–(71) we get,

$$\iiint_V \left(\phi\nabla^2\psi - \psi\nabla^2\phi\right) dv = \iint_S (\phi\nabla\psi - \psi\nabla\phi)\cdot n\, dS$$

This is called the Green's second theorem (or identity) Hence, the Green's theorem can also be stated as

$$\iiint_V \left(\phi\nabla^2\psi - \psi\nabla^2\phi\right) dv = \iint_S \left(\phi\frac{\partial\psi}{\partial n} - \psi\frac{\partial\phi}{\partial n}\right) dS$$

Where $\frac{\partial\psi}{\partial n}$ and $\frac{\partial\phi}{\partial n}$ denote the directional derivative of ϕ and ψ respectively along the outward drawn normal at any point of *S*.

Other forms of Green's Theorem

Green's theorem is a vector identity which is equivalent

to the curl theorem in the plane. Over a region D in the plane with boundary ∂D, Green's theorem states

$$\int_{\partial D}[f(x,y)dx+g(x,y)dy]=\iint_D\left(\frac{\partial g}{\partial x}-\frac{\partial f}{\partial y}\right)dxdy$$

where the left side is a line integral and the right side is a surface integral. This can also be written compactly in vector form as,

$$\int_{\partial D}F\cdot dS=\iint_D(\nabla\times F)\cdot \mathrm{da}$$

If the region D is on the left when traveling around ∂D, then area of D can be computed using the elegant formula

$$\mathrm{A}=\frac{1}{2}\int_{\partial D}xdy-ydx$$

giving a surprising connection between the area of a region and the line integral around its boundary. For a plane curve specified parametrically as $(x(t), y(t))$, equation becomes

$$A=\frac{1}{2}\int\left(x\frac{dy}{dt}-y\frac{dx}{dt}\right)dt$$

which gives the *signed* area enclosed by the curve.

GREEN'S THEOREM FOR AREA IN PLANE ELATING TO ITS BOUNDING CURVE

The analog of the *"Fundamental Theorem of Differential and Integral Calculus"* for a region Â bounded in a plane with normal k that is bounded by a curve ∂ Â is:

$$\iint_R\left(\nabla\times\vec{F}\right)\cdot\hat{k}\;dxdy=\oint_{\partial R}\vec{F}\cdot d\vec{r}$$

Within the domain, the rotational flow (Ñ ′ F) from one cell moves into its neigthbours; however, at the edges the local rotation is a net loss or gain. The local net loss or gain is $\vec{F}\cdot(dx,dy)$. The generalization of this idea to a surface $\partial\mathfrak{R}$ bounding a domain $\mathfrak{R}$ results in Stokes' theorem.

REDUCTION OF SURFACE INTEGRAL TO LINE INTEGRAL

Let the region R on the surface be break up into tiny pieces each one looking planar, apply Green's theorem on each and add up the mess to get the same theorem on all of R. The result is called *Stokes' Theorem*. Theorem: With R an arbitrary reasonably smooth and nice region on a surface that is locally planar, and with δR its boundary, we have

$$\iint_R \overline{\nabla} \times \overline{vdS} = \oint_{\delta R} \nabla \cdot d\bar{l}$$

We can use it as a tool for rewriting integrals of one kind as integrals of another and for proving peculiar facts about irrotational fields (those for which curl = 0). It can be stated in the following form:The circulation integral of v over a bounding curve is the flux integral of the curl of v over any surface it bounds. Notice that this theorem immediately implies that if R is simply connected and curl $v = 0$ throughout it, then the potential f (with v = Ñf) defined as path integral is path independent.

Stokes' Theorem

The divergence theorem is used to find a surface integral over a closed surface and Green's theorem is use to find a line integral that encloses a surface (region) in the xy-plane. This theorem relates the surface integral to a line integral. Let S be a oriented surface with unit normal vector N and let C be the boundary of S. Then C is *positively oriented* if its orientation follows the right hand rule, that is if you right hand curls around N in the direction of C's orientation, then your thumb will be pointing in the direction of N.

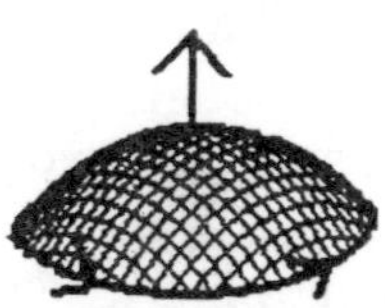

Statement: Let S be an oriented surface with unit normal vector n and C be the positively oriented boundary of S. If F is a vector field with continuous first order partial derivatives then,

$$\int_C F \cdot dr = \iint_S (CurlF) \cdot ndS$$

In other words: The line integral of the tangential component of a vector f taken around a simple closed curve C is equal to the surface integral of the normal component of the curl of f taken over any surface *S* having C as its boundary.

Cartesian form of Stokes' Theorem for plane

Let us take a system of Cartesian rectangular coordinate axes such that the plane of the given surface *S* is the *xy*-plane and z-axis lies along the direction of the normal vector *n*. Here, the normal vector is constant

Let $F = F_i i + F_2 j + F_3 k$

$$\therefore F \cdot dr = (F_i i + F_2 j + F_3 k) \cdot (dxi + dyj + dzk)$$

$$\Rightarrow F \cdot dr = F_i dx + F_2 dy + F_3 dz$$

Now,

$$\iint_S (Curl F) \cdot n dS = \iint_S (Curl F) \cdot k dS$$

$$= \iint_S \left(\frac{\partial F_2}{\partial x} - \frac{\partial F_1}{\partial y} \right) dxdy$$

$\therefore$ Stokes' Theorem for plane becomes

$$\oint_C (F_i dx + F_2 dy) = \iint_S \left(\frac{\partial F_2}{\partial x} - \frac{\partial F_1}{\partial y} \right) dxdy$$

Let us, now prove it. Let the region *S* be such that any line parallel to either axis meet the boundary C in at the most two points. Let the region be $x = a$, $y = \phi_1(x)$, $y = \phi_2(x)$, where $\phi_2(x)$ 3 $\phi_1(x)$.

The boundary C is split into two arc C_1 and C_2 as shown.

$$\therefore \iint_S \frac{\partial F_1}{\partial y} dxdy = \int_a^b \left[\int_{\phi_1}^{\phi_2} \frac{\partial F_1}{\partial y} dy \right] dx$$

$$= \int_a^b \left[F_1(x, \phi_2(x)) - F_1(x, \phi_1(x)) \right] dx$$

$$= -\iint_{C_2} F_1(x, y) dx - \iint_{C_1} F_1(x, y) dx$$

$$= -\oint_C F_1(x, y) dx$$

Similarly,

$$\therefore \iint_S \frac{\partial F_2}{\partial y}\, dxdy = -\oint_C F_2(x,y)\,dx$$

$\therefore$ From (82) and (83), we get,

$$\oint_C (F_i dx + F_2 dy) = \iint_S \left(\frac{\partial F_2}{\partial x} - \frac{\partial F_1}{\partial y}\right) dxdy$$

Hence the result. If the region S is such that a line parallel to either axes meets the boundary C in more than tow points, we divide S into a finite number of sub-regions such that the boundary of each is met in at the most two points by any line parallel to either axis. By applying the above result to each sub-region and then adding the results, we find that the theorem is true for S. This because the line integrals along the boundary curves will cancel in pairs.

Cartesian form of Stokes' Theorem in space

Let $f = f_1 i + f_2 j + f_3 k$. Let the outward drawn normal vector n makes angles α, β, γ with the positive direction of x, y, z axes respectively so that,

$n = \cos\alpha i + \cos\beta j + \cos\gamma k$

$$\text{Then, } \nabla' f = \begin{vmatrix} i & j & k \\ \frac{\partial}{\partial x} & \frac{\partial}{\partial y} & \frac{\partial}{\partial z} \\ f_1 & f_2 & f_3 \end{vmatrix}$$

$$= \left[i\left(\frac{\partial f_3}{\partial x} - \frac{\partial f_2}{\partial z}\right) + j\left(\frac{\partial f_1}{\partial z} - \frac{\partial f_3}{\partial x}\right) + k\left(\frac{\partial f_2}{\partial x} - \frac{\partial f_1}{\partial y}\right)\right]$$

$$(\nabla \times f)\cdot n = \left[i\left(\frac{\partial f_3}{\partial x} - \frac{\partial f_2}{\partial z}\right) + j\left(\frac{\partial f_1}{\partial z} - \frac{\partial f_3}{\partial x}\right) + k\left(\frac{\partial f_2}{\partial x} - \frac{\partial f_1}{\partial y}\right)\right] \cdot (i\cos\alpha + j\cos\beta + k\cos\gamma)$$

$$= \left[\left(\frac{\partial f_3}{\partial x} - \frac{\partial f_2}{\partial z}\right)\cos\alpha + \left(\frac{\partial f_1}{\partial z} - \frac{\partial f_3}{\partial x}\right)\cos\beta + \left(\frac{\partial f_2}{\partial x} - \frac{\partial f_1}{\partial y}\right)\cos\gamma\right]$$

Also, $f \cdot dr = (f_1 i + f_2 j + f_3 k)\cdot(dxi + dyj + dzk)$

$= f_1 dx + f_2 dy + f_3 dz$

∴ Stokes' Theorem becomes

$$\int_C (f_1 dx + f_2 dy + f_3 dz)$$

$$\iint \left[\left(\frac{\partial f_3}{\partial x} - \frac{\partial f_2}{\partial z}\right)\cos\alpha + \left(\frac{\partial f_1}{\partial z} - \frac{\partial f_3}{\partial x}\right)\cos\beta + \left(\frac{\partial f_2}{\partial x} - \frac{\partial f_1}{\partial y}\right)\cos\gamma\right] dS$$

Curl and Circulation

Just as the divergence theorem assisted us in understanding the divergence of a function at a point, Stokes' theorem helps us understand what the Curl of a vector field is. Let P be a point on the surface and C_e be a tiny circle around P on the surface.

$$\int_C F.dr$$

The measures the amount of circulation around p. You can see this by noticing that if F flows in the direction of the tangent vector, then $F \cdot dr$ will be positive. If it flows in the opposite direction, then it will be negative. The stronger the force field in the direction of the tangent vector, the greater the circulation. Since the region enclosed by C_e is tiny, the surface integral can be approximated by,

$$\iint_S (CurlF \cdot n) dS \approx CurlF(P) \cdot n(\text{Area of C})$$

or,

Curl F . n = Circulation per unit area.

5

Statistics

STATISTICS

Statistics is a mathematical science pertaining to the collection, analysis, interpretation or explanation, and presentation of data. It is applicable to a wide variety of academic disciplines, from the physical and social sciences to the humanities. Statistics are also used for making informed decisions. Statistical methods can be used to summarize or describe a collection of data; this is called descriptive statistics. In addition, patterns in the data may be modeled in a way that accounts for randomness and uncertainty in the observations, and then used to draw inferences about the process or population being studied; this is called inferential statistics. Both descriptive and inferential statistics comprise applied statistics. There is also a discipline called mathematical statistics, which is concerned with the theoretical basis of the subject. The idea that statistics branched off from mathematics is a widely held misconception. Some place an undue emphasis on the relationship, but the two disciplines are very different.

The purpose of descriptive statistics is to communicate information, while inferential statistics is used to reach conclusions and deductions that possibly explain the data. Both of these together make up applied statistics. There is also a discipline called mathematical statistics, which is concerned with the theoretical basis of the subject.

Mathematical Statistics

Mathematical statistics uses probability theory and other

branches of mathematics to study statistics from a purely mathematical standpoint. Mathematical statistics deals with gaining information from data. In practice, data often contain some randomness or uncertainty. Statistics handles such data using methods of probability theory.

Statistics is divided into:

- Descriptive statistics-the part of mathematical statistics that describes data, *i.e.* summarises the data and their typical properties
- Inferential statistics-the part of mathematical statistics that draws conclusions from data, *i.e.* checks whether the data fulfill some condition and gives guarantees on the involved uncertainty.

Mathematical statistics is the theoretical basis for many practices in applied statistics.

Descriptive Statistics

Descriptive Statistics are used to describe the basic features of the data in a study. They provide simple summaries about the sample and the measures. Together with simple graphics analysis, they form the basis of virtually every quantitative analysis of data.

Various techniques that are commonly used are classified as:

1. Graphical description in which we use graphs to summarize data.
2. Tabular description in which we use tables to summarize data.
3. Summary statistics in which we calculate certain values to summarize data.

In general, statistical data can be described as a list of *subjects* or *units* and the data associated with each of them. Although most research uses many data types for each *unit*, we will limit ourselves to just one data item each for this simple introduction.

We have two objectives for our summary:

1. We want to choose a statistic that shows how different *units* seem similar. Statistical textbooks call the solution to this objective, a *measure of central tendency*.

2. We want to choose another statistic that shows how they differ. This kind of statistic is often called a *measure of statistical variability.*

When we are summarizing a quantity like length or weight or age, it is common to answer the first question with the arithmetic mean, the median, or the mode. Sometimes, we choose specific values from the cumulative distribution function called quantiles. The most common measures of variability for quantitative data are the variance; its square root, the standard deviation; the range; interquartile range; and the average absolute deviation (average deviation).

One important use of descriptive statistics is to summarize a collection of data in a clear and understandable way. For example, assume a psychologist gave a personality test measuring shyness to all 2500 students attending a small college. How might these measurements bė summarized? There are two basic methods: numerical and graphical. Using the numerical approach one might compute statistics such as the mean and standard deviation. These statistics convey information about the average degree of shyness and the degree to which people differ in shyness. Using the graphical approach one might create a stem and leaf display and a box plot. These plots contain detailed information about the distribution of shyness scores.

Graphical methods are better suited than numerical methods for identifying patterns in the data. Numerical approaches are more precise and objective. Since the numerical and graphical approaches compliment each other, it is wise to use both.

Statistical Inference

Inferential statistics or statistical induction comprises the use of statistics to make inferences concerning some unknown aspect of a population. It is distinguished from descriptive statistics.

Statistical inference concerns the problem of inferring properties of an unknown distribution from data generated by that distribution. The most common type of inference involves approximating the unknown distribution by choosing a distribution from a restricted family of distributions. Generally the restricted family of distributions is specified parametrically.

CONCEPTS IN STATISTICS

In this introduction, we will briefly discuss those elementary statistical concepts that provide the necessary foundations for more specialized expertise in any area of statistical data analysis. The selected topics illustrate the basic assumptions of most statistical methods and/or have been demonstrated in research to be necessary components of one's general understanding of the "quantitative nature" of reality.

Because of space limitations, we will focus mostly on the functional aspects of the concepts discussed and the presentation will be very short. Further information on each of those concepts can be found in statistical textbooks.

VARIABLES

In statistics, variables refer to measurable attributes, as these typically vary over time or between individuals. Variables can be discrete (taking values from a finite or countable set), continuous (having a continuous distribution function), or neither. Temperature is a continuous variable, while the number of legs of an animal is a discrete variable. This concept of a variable is widely used in the natural, medical and social sciences.

In causal models, a distinction is made between *"independent variables"* and *"dependent variables"*, the latter being expected to vary in value in response to changes in the former. In other words, an independent variable is presumed to potentially affect a dependent one. In experiments, independent variables include factors that can be altered or chosen by the researcher independent of other factors.

For example, in an experiment to test whether or not the boiling point of water changes with altitude, the altitude is under direct control and is the *independent* variable, and the boiling point is presumed to depend upon it and is therefore the *dependent* variable. The collection of results from an experiment, or information to be used to draw conclusions, is known as data. It is often important to consider which variables to allow for, or to directly control or eliminate, in the design of experiments.

While independent variables can refer to quantities and qualities that are under experimental control, they can also include extraneous factors that influence results in a confusing or undesired manner. In general, if strongly confounding variables exist that can substantially affect the result, then this makes it more difficult to interpret the results.

For example, a study into the incidence of cancer with age will also have to take into account *variables* such as income (poorer people may have less healthy lives), location (some cancers vary depending on diet and sunlight), stress and lifestyle issues (cancer may be related to these more than age), and so on. Failure to at least consider these factors can lead to grossly inaccurate deductions. For this reason, controlling unwanted variables is important in research.

Variables are things that we measure, control, or manipulate in research. They differ in many respects, most notably in the role they are given in our research and in the type of measures that can be applied to them. A symbol or name that stands for a value.

For example, in the expression $x + y$ x and y are variables.

Measurement Scales

Variables differ in "how well" they can be measured, *i.e.*, in how much measurable information their measurement scale can provide. There is obviously some measurement error involved in every measurement, which determines the "amount of information" that we can obtain. Another factor that determines the amount of information that can be provided by a variable is its "type of measurement scale." Specifically variables are classified as (a) nominal, (b) ordinal, (c) interval or (d) ratio.

1. Nominal variables allow for only qualitative classification. That is, they can be measured only in

terms of whether the individual items belong to some distinctively different categories, but we cannot quantify or even rank order those categories. For example, all we can say is that 2 individuals are different in terms of variable A (*e.g.*, they are of different race), but we cannot say which one "has more" of the quality represented by the variable. Typical examples of nominal variables are gender, race, colour, city, etc.

2. Ordinal variables allow us to rank order the items we measure in terms of which has less and which has more of the quality represented by the variable, but still they do not allow us to say "how much more." A typical example of an ordinal variable is the socioeconomic status of families. For example, we know that upper-middle is higher than middle but we cannot say that it is, for example, 18% higher. Also this very distinction between nominal, ordinal, and interval scales itself represents a good example of an ordinal variable. For example, we can say that nominal measurement provides less information than ordinal measurement, but we cannot say "how much less" or how this difference compares to the difference between ordinal and interval scales.
3. Interval variables allow us not only to rank order the items that are measured, but also to quantify and compare the sizes of differences between them. For example, temperature, as measured in degrees Fahrenheit or Celsius, constitutes an interval scale. We can say that a temperature of 40 degrees is higher than a temperature of 30 degrees, and that an increase from 20 to 40 degrees is twice as much as an increase from 30 to 40 degrees.
4. Ratio variables are very similar to interval variables; in addition to all the properties of interval variables, they feature an identifiable absolute zero point, thus they allow for statements such as x is two times more than y. Typical examples of ratio scales are

measures of time or space. For example, as the Kelvin temperature scale is a ratio scale, not only can we say that a temperature of 200 degrees is higher than one of 100 degrees, we can correctly state that it is twice as high. Interval scales do not have the ratio property. Most statistical data analysis procedures do not distinguish between the interval and ratio properties of the measurement scales.

Dependent and Independent Variables

Independent variables are those that are manipulated whereas dependent variables are only measured or registered. This distinction appears terminologically confusing to many because, as some students say, "all variables depend on something." However, once you get used to this distinction, it becomes indispensable. The terms dependent and independent variable apply mostly to experimental research where some variables are manipulated, and in this sense they are "independent" from the initial reaction patterns, features, intentions, etc. of the subjects.

Some other variables are expected to be "dependent" on the manipulation or experimental conditions. That is to say, they depend on "what the subject will do" in response. Somewhat contrary to the nature of this distinction, these terms are also used in studies where we do not literally manipulate independent variables, but only assign subjects to "experimental groups" based on some pre-existing properties of the subjects. For example, if in an experiment, males are compared with females regarding their white cell count (WCC), Gender could be called the independent variable and WCC the dependent variable.

Correlational and Experimental Research

Most empirical research belongs clearly to one of those two general categories. In correlational research we do not (or at least try not to) influence any variables but only measure them and look for relations (correlations) between some set of variables, such as blood pressure and cholesterol level. In experimental research, we manipulate some variables and then measure the effects of this manipulation on other variables; for example, a researcher might artificially increase blood pressure and then record cholesterol level. Data analysis in experimental research also comes down to calculating "correlations" between variables, specifically, those manipulated and those affected by the manipulation. However, experimental data may potentially provide qualitatively better information: Only experimental data can conclusively demonstrate causal relations between variables. For example, if we found that whenever we change variable A then variable B changes, then we can conclude that "A influences B." Data from correlational research can only be "interpreted" in causal terms based on some theories that we have, but correlational data cannot conclusively prove causality.

VARIABLES AND ITS RELATIONS

Regardless of their type, two or more variables are related if in a sample of observations, the values of those variables are distributed in a consistent manner. In other words, variables are related if their values systematically correspond to each other for these observations. For example, Gender and WCC would be considered to be related if most males had high WCC and most females low WCC, or vice versa; Height is related to Weight because typically tall individuals are heavier than short ones; IQ is related to the Number of Errors in a test, if people with higher IQ's make fewer errors.

Generally speaking, the ultimate goal of every research or scientific analysis is finding relations between variables. The philosophy of science teaches us that there is no other way of representing "meaning" except in terms of relations between some quantities or qualities; either way involves relations between variables. Thus, the advancement of science must always involve finding new relations between variables. Correlational research involves measuring such relations in the most straightforward manner. However, experimental research is not any different in this respect. For example, the above mentioned experiment comparing WCC in males and females can be described as looking for a correlation between two variables: Gender and WCC. Statistics does nothing else but help us evaluate relations between variables. Actually, all of the hundreds of procedures that are described in this manual can be interpreted in terms of evaluating various kinds of inter-variable relations.

FEATURES OF VARIABLES

The two most elementary formal properties of every relation between variables are the relation's (a) magnitude (or "size") and (b) its reliability (or "truthfulness").

1. Magnitude (or "size"). The magnitude is much easier to understand and measure than reliability. For example, if every male in our sample was found to have a higher WCC than any female in the sample, we could say that the magnitude of the relation between the two variables (Gender and WCC) is very high in our sample. In other words, we could predict one based on the other (at least among the members of our sample).
2. Reliability (or "truthfulness"). The reliability of a relation is a much less intuitive concept, but still extremely important. It pertains to the "representativeness" of the result found in our specific sample for the entire population. In other words, it says how probable it is that a similar relation would be found if the experiment was replicated with other samples drawn from the same population. Remember that we are almost never "ultimately" interested only in what is going on in our sample; we are interested in the sample only to the extent it can provide information about the population. If our study meets some specific criteria (to be mentioned later), then the reliability of a relation between variables observed in our sample can be quantitatively estimated and represented using a standard measure (technically called p-value or statistical significance level.

STATISTICAL SIGNIFICANCE

The statistical significance of a result is the probability that the observed relationship (*e.g.*, between variables) or a difference (*e.g.*, between means) in a sample occurred by pure chance ("luck of the draw"), and that in the population from which the sample was drawn, no such relationship or differences exist. Using less technical terms, one could say that the statistical significance of a result tells us something about the degree to which the result is "true" (in the sense of being "representative of the population").

More technically, the value of the p-value represents a decreasing index of the reliability of a result. The higher the p-value, the less we can believe that the observed relation between variables in the sample is a reliable indicator of the relation between the respective variables in the population. Specifically, the p-value represents the probability of error that is involved in accepting our observed result as valid, that is, as "representative of the population." For example, a p-value of.05 (*i.e.*,1/20) indicates that there is a 5% probability that the relation between the variables found in our sample is a "fluke."

In other words, assuming that in the population there was no relation between those variables whatsoever, and we were repeating experiments like ours one after another, we could expect that approximately in every 20 replications of the experiment there would be one in which the relation between the variables in question would be equal or stronger than in ours. Note that this is not the same as saying that, given that there IS a relationship between the variables, we can expect to replicate the results 5% of the time or 95% of the time; when there is a relationship between the variables in the population, the probability of replicating the study and finding that relationship is related to the statistical power of the design. In many areas of research, the p-value of.05 is customarily treated as a "border-line acceptable" error level.

DETERMINATION OF SIGNIFICANT RESULT

There is no way to avoid arbitrariness in the final decision as to what level of significance will be treated as really "significant." That is, the selection of some level of significance, up to which the results will be rejected as invalid, is arbitrary. In practice, the final decision usually depends on whether the outcome was predicted a priori or only found post hoc in the course of many analyses and comparisons performed on the data set, on the total amount of consistent supportive evidence in the entire data set, and on "traditions" existing in the particular area of research.

Typically, in many sciences, results that yield $p \leq .05$ are considered borderline statistically significant but remember that this level of significance still involves a pretty high probability of error (5%). Results that are significant at the $p \leq .01$ level are commonly considered statistically significant, and $p \leq .005$ or $p \leq .001$ levels are often called "highly" significant. But remember that those classifications represent nothing else but arbitrary conventions that are only informally based on general research experience. Needless to say, the more analyses you perform on a data set, the more results will meet "by chance" the conventional significance level.

For example, if you calculate correlations between ten variables (*i.e.*, 45 different correlation coefficients), then you should expect to find by chance that about two (*i.e.*, one in every 20) correlation coefficients are significant at the $p \leq .05$ level, even if the values of the variables were totally random and those variables do not correlate in the population.

Some statistical methods that involve many comparisons, and thus a good chance for such errors, include some "correction" or adjustment for the total number of comparisons. However, many statistical methods (especially simple exploratory data analyses) do not offer any straightforward remedies to this problem. Therefore, it is up to the researcher to carefully evaluate the reliability of unexpected findings. Many examples in this manual offer specific advice on how to do this; relevant information can also be found in most research methods textbooks.

We said before that strength and reliability are two different features of relationships between variables. However, they are not totally independent. In general, in a sample of a particular size, the larger the magnitude of the relation between variables, the more reliable the relation.

Assuming that there is no relation between the respective variables in the population, the most likely outcome would be also finding no relation between those variables in the research sample. Thus, the stronger the relation found in the sample, the less likely it is that there is no corresponding relation in the population. As we see, the magnitude and significance of a relation appear to be closely related, and we could calculate the significance from the magnitude and vice-versa; however, this is true only if the sample size is kept constant, because the relation of a given strength could be either highly significant or not significant at all, depending on the sample size.

If there are very few observations, then there are also respectively few possible combinations of the values of the variables, and thus the probability of obtaining by chance a combination of those values indicative of a strong relation is relatively high. Consider the following illustration. If we are interested in two variables (Gender: male/female and WCC: high/low) and there are only four subjects in our sample (two males and two females), then the probability that we will find, purely by chance, a 100% relation between the two variables can be as high as one-eighth.

Specifically, there is a one-in-eight chance that both males will have a high WCC and both females a low WCC, or vice versa. Now consider the probability of obtaining such a perfect match by chance if our sample consisted of 100 subjects; the probability of obtaining such an outcome by chance would be practically zero. Let's look at a more general example. Imagine a theoretical population in which the average value of WCC in males and females is exactly the same.

Needless to say, if we start replicating a simple experiment by drawing pairs of samples (of males and females) of a particular size from this population and calculating the difference between the average WCC in each pair of samples, most of the experiments will yield results close to 0. However, from time to time, a pair of samples will be drawn where the difference between males and females will be quite different from 0. How often will it happen?

The smaller the sample size in each experiment, the more likely it is that we will obtain such erroneous results, which in this case would be results indicative of the existence of a relation between gender and WCC obtained from a population in which such a relation does not exist.

Example. "Baby Boys to Baby Girls Ratio"

Consider the following example from research on statistical reasoning. There are two hospitals: in the first one, 120 babies are born every day, in the other, only 12. On average, the ratio of baby boys to baby girls born every day in each hospital is 50/50. However, one day, in one of those hospitals twice as many baby girls were born as baby boys. In which hospital was it more likely to happen? The answer is obvious for a statistician, but as research shows, not so obvious for a lay person: It is much more likely to happen in the small hospital. The reason for this is that technically speaking, the probability of a random deviation of a particular size (from the population mean), decreases with the increase in the sample size.

Significant Result

The smaller the relation between variables, the larger the sample size that is necessary to prove it significant. For example, imagine how many tosses would be necessary to prove that a coin is asymmetrical if its bias were only.000001%! Thus, the necessary minimum sample size increases as the magnitude of the effect to be demonstrated decreases. When the magnitude of the effect approaches 0, the necessary sample size to conclusively prove it approaches infinity.

That is to say, if there is almost no relation between two variables, then the sample size must be almost equal to the population size, which is assumed to be infinitely large. Statistical significance represents the probability that a similar outcome would be obtained if we tested the entire population.

Thus, everything that would be found after testing the entire population would be, by definition, significant at the highest possible level, and this also includes all "no relation" results.

MAGNITUDE OF RELATIONS

There are very many measures of the magnitude of relationships between variables which have been developed by statisticians; the choice of a specific measure in given circumstances depends on the number of variables involved, measurement scales used, nature of the relations, etc. Almost all of them, however, follow one general principle: they attempt to somehow evaluate the observed relation by comparing it to the "maximum imaginable relation" between those specific variables. Technically speaking, a common way to perform such evaluations is to look at how differentiated are the values of the variables, and then calculate what part of this "overall available differentiation" is accounted for by instances when that differentiation is "common" in the two (or more) variables in question. Speaking less technically, we compare "what is common in those variables" to "what potentially could have been common if the variables were perfectly related."

Let us consider a simple illustration. Let us say that in our sample, the average index of WCC is 100 in males and 102 in females. Thus, we could say that on average, the deviation of each individual score from the grand mean (101) contains a component due to the gender of the subject; the size of this component is 1. That value, in a sense, represents some measure of relation between Gender and WCC. However, this value is a very poor measure, because it does not tell us how relatively large this component is, given the "overall differentiation" of WCC scores. Consider two extreme possibilities:

1. If all WCC scores of males were equal exactly to 100, and those of females equal to 102, then all deviations from the grand mean in our sample would be entirely accounted for by gender. We would say that in our sample, gender is perfectly correlated with WCC, that is, 100% of the observed differences between subjects

regarding their WCC is accounted for by their gender.

2. If WCC scores were in the range of 0-1000, the same difference (of 2) between the average WCC of males and females found in the study would account for such a small part of the overall differentiation of scores that most likely it would be considered negligible. For example, one more subject taken into account could change, or even reverse the direction of the difference. Therefore, every good measure of relations between variables must take into account the overall differentiation of individual scores in the sample and evaluate the relation in terms of (relatively) how much of this differentiation is accounted for by the relation in question.

Statistical Tests

Because the ultimate goal of most statistical tests is to evaluate relations between variables, most statistical tests follow the general format that was explained in the previous paragraph. Technically speaking, they represent a ratio of some measure of the differentiation common in the variables in question to the overall differentiation of those variables. For example, they represent a ratio of the part of the overall differentiation of the WCC scores that can be accounted for by gender to the overall differentiation of the WCC scores. This ratio is usually called a ratio of explained variation to total variation. In statistics, the term explained variation does not necessarily imply that we "conceptually understand" it. It is used only to denote the common variation in the variables in question, that is, the part of variation in one variable that is "explained" by the specific values of the other variable, and vice versa.

Level of Statistical Significance

Let us assume that we have already calculated a measure of a relation between two variables (as explained above). The next question is "how significant is this relation?" For example, is 40% of the explained variance between the two variables enough to consider the relation significant? The answer is "it depends." Specifically, the significance depends mostly on the sample size. As explained before, in very large samples, even very small relations between variables will be significant, whereas in very small samples even very large relations cannot be considered reliable (significant).

Thus, in order to determine the level of statistical significance, we need a function that represents the relationship between "magnitude" and "significance" of relations between two variables, depending on the sample size. The function we need would tell us exactly "how likely it is to obtain a relation of a given magnitude (or larger) from a sample of a given size, assuming that there is no such relation between those variables in the population." In other words, that function would give us the significance (p) level, and it would tell us the probability of error involved in rejecting the idea that the relation in question does not exist in the population.

This "alternative" hypothesis (that there is no relation in the population) is usually called the null hypothesis. It would be ideal if the probability function was linear, and for example, only had different slopes for different sample sizes. Unfortunately, the function is more complex, and is not always exactly the same; however, in most cases we know its shape and can use it to determine the significance levels for our findings in samples of a particular size. Most of those functions are related to a general type of function which is called normal.

Normal Distribution

Where pairs of samples of males and females were drawn from a population in which the average value of WCC in males and females was exactly the same. Although the most likely outcome of such experiments (one pair of samples per experiment) was that the difference between the average WCC in males and females in each pair is close to zero, from time to time, a pair of samples will be drawn where the difference between males and females is quite different from 0.

How often does it happen? If the sample size is large enough, the results of such replications are "normally distributed" (this important principle is explained and illustrated in the next paragraph), and thus knowing the shape of the normal curve, we can precisely calculate the probability of obtaining "by chance" outcomes representing various levels of deviation from the hypothetical population mean of 0.

If such a calculated probability is so low that it meets the previously accepted criterion of statistical significance, then we have only one choice: conclude that our result gives a better approximation of what is going on in the population than the "null hypothesis" (remember that the null hypothesis was considered only for "technical reasons" as a benchmark against which our empirical result was evaluated). Note that this entire reasoning is based on the assumption that the shape of the distribution of those "replications" (technically, the "sampling distribution") is normal. This assumption is discussed in the next paragraph.

Not all, but most of them are either based on the normal distribution directly or on distributions that are related to, and can be derived from normal, such as t, F, or Chi-square. Typically, those tests require that the variables analysed are themselves normally distributed in the population, that is, they meet the so-called "normality assumption." Many observed variables actually are normally distributed, which is another reason why the normal distribution represents a "general feature" of empirical reality.

The problem may occur when one tries to use a normal distribution-based test to analyse data from variables that are themselves not normally distributed. In such cases we have two general choices. First, we can use some alternative "nonparametric" test; but this is often inconvenient because such tests are typically less powerful and less flexible in terms of types of conclusions that they can provide. Alternatively, in many cases we can still use the normal distribution-based test if we only make sure that the size of our samples is large enough. The latter option is based on an extremely important principle which is largely responsible for the popularity of tests that are based on the normal function. Namely, as the sample size increases, the shape of the sampling distribution (*i.e.,* distribution of a statistic from the sample; this term was first used by Fisher, 1928a) approaches normal shape, even if the distribution of the variable in question is not normal. This principle is illustrated in the following animation showing a series of sampling distributions (created with gradually increasing sample sizes of: 2, 5, 10, 15, and 30) using a variable that is clearly non-normal in the population, that is, the distribution of its values is clearly skewed.

However, as the sample size (of samples used to create the sampling distribution of the mean) increases, the shape of the sampling distribution becomes normal. Note that for n=30, the shape of that distribution is "almost" perfectly normal. This principle is called the central limit theorem.

USES OF STATISTICS

One of the potential shortfalls of anecdotal data is that they are idiosyncratic. Just as the congressional staffer told me her father received a better education from the high school they both attended than she did, We could have easily received a higher quality education than my father did. Statistics allow researchers to collect information, or data, from a large number of people and then summarize their typical experience. Do *most* people receive a better or worse education than their parents?

Statistics allow researchers to take a large batch of data and *summarize* it into a couple of numbers, such as an average. Of course, when a large amount of data is summarized into a single number, a lot of information is lost, including the fact that different people have very different experiences. So it is important to remember that, for the most part, statistics do not provide useful information about each individual's experience. Rather, researchers generally use statistics to make *general* statements about a population. Although personal stories are often moving or interesting, it is often important to understand what the *typical* or *average* experience is. For this, we need statistics. Statistics are also used to reach conclusions about general differences between groups. For example, suppose that in our family, there are four children, two men and two women. Suppose that the women in our family are taller than the men. This personal experience may lead me to the conclusion that women are generally taller than men. Of course, we know that, on average, men are taller than women.

The reason we know this is because researchers have taken large, random samples of men and women and compared their average heights. Researchers are often interested in making such comparisons: Do cancer patients survive longer using one drug than another? Is one method of teaching children to read more effective than another? Do men and women differ in their enjoyment of a certain movie? To answer these questions, we need to collect data from randomly selected samples and compare these data using statistics. The results we get from such comparisons are often more trustworthy than the simple observations people make from non-random samples, such as the different heights of men and women in our family.

Statistics can also be used to see if scores on two variables are related and to make predictions. For example, statistics can be used to see whether smoking cigarettes is related to the likelihood of developing lung cancer. For years, tobacco companies argued that there was no relationship between smoking and cancer. Sure, some people who smoked developed cancer.

But the tobacco companies argued that (a) many people who smoked never developed cancer, and (b) many people who smoke tend to do other things that may lead to cancer development, such as eating unhealthy foods and not exercising. With the help of statistics in a number of studies, researchers were finally able to produce a preponderance of evidence indicating that, in fact, there is a relationship between cigarette smoking and cancer. Because statistics tend to focus on overall patterns rather than individual cases, this research did not suggest that *everyone* who smokes would develop cancer.

Rather, the research demonstrated that, on average, people have a greater chance of developing cancer if they smoke cigarettes than if they do not with a moment's thought, you can imagine a large number of interesting and important questions that statistics about relationships can help you answer. Is there a relationship between self-esteem and academic achievement? Is there a relationship between the appearance of criminal defendants and their likelihood of being convicted? Is it possible to predict the violent crime rate of a state from the amount of money the state spends on drug treatment programmes? If we know the fathers' height, how accurately can we predict sons' height? These and thousands of other questions have been examined by researchers using statistics designed to determine the relationship between variables in a population.

6

Data Handling

TYPES OF DATA

There are two types of data that can be manipulated with statistical tools: cross-sectional data and time-series data. *Cross-sectional data* is a snapshot that is collected at a single point in time. It has no time dimension. A typical example of cross-sectional data would be a set of height measurements of children of different ages. Computing the average of all the measurements for a given age gives you the average height of a child of that age. The specific date and time on which any given measurement was taken is irrelevant. All that matters is the height of each child and the child's age when that height was measured.

The second type of data that you'll encounter is time-series data. With *time-series data,* every measurement has a time dimension associated with it. For example, you can look at stock-market prices over some period of time, and you'll find that the share price for any given company differs from day to day. And, not only from day to day, but also typically from hour to hour and minute to minute. A share price by itself means nothing. It's the share price coupled with the time at which that price was effective that is meaningful. A list of such prices and times over a given period is referred to as a time series.

We need to define five more terms before proceeding into the sections that describe the calculations behind the various statistics discussed in this chapter. These terms are population, sample, sample size, case, and value.

Consider the problem of having a large box of oranges for which you need to determine the sweetness level so that you can sell the box at the best possible price. That box of oranges represents your *population*. Now you could taste each orange in the box to check the sweetness level, but then you wouldn't have any oranges left to sell. Instead, you might choose to test only a few oranges—say, three. Those oranges that you test represent a *sample*, and the number of oranges in the sample represents the *sample size*. From a statistical point of view, you are testing a representative sample and then extrapolating the results to the oranges remaining in the box.

Continuing with our example, each orange in a sample represents a specific *case*. For each case—or orange—you must test the sweetness level. The result of each test will be a *value*—usually a number—indicating how sweet the orange really is. Throughout this chapter, we use the term *case* to refer to a specific measurement in the sample. We use the term *value* to refer to a distinct numeric result. A value may occur more than once in a sample. For example, three oranges (three cases) may all have the same sweetness value.

Now that you have a good grasp of the statistical terms used in this chapter, you can proceed to the following sections describing various statistical measurements and calculations. The calculations described in these upcoming sections form the basis for the recipes shown later in the chapter.

Whenever you collect data, you end up with a group of scores on one or more variables. If you take the scores on one variable and arrange them in order from lowest to highest, what you get is a distribution of scores. Researchers often want to know about the characteristics of these distributions of scores, such as the shape of the distribution, how spread out the scores are, what the most common score is, and so on. One set of distribution characteristics that researchers are usually interested in is central tendency. This set consists of the mean, median, and mode.

USE OF SPECIAL SYMBOLS

In this chapter, we use some mathematical symbols that

you may not have seen used since you were in college. In this sidebar, we present a short review of these symbols to help you better understand our recipes.

The Σ symbol is the summation operator. It's used as a shorthand way of telling you to add up all the values in a sample set. When you see ΣX in a formula, you know that you should sum all the values in the sample set labeled *X*. Some formulas are more complex than that. For example, you might see ΣXY in a formula.

The *X* and *Y* refer to two sample sets, and this notation is telling you to multiply corresponding elements of each set together and sum the results. So, multiply the first case in *X* by the first case in *Y*.

Then, add to that the result of multiplying the second case in *X* by the second case in *Y*. Continue this pattern for all cases in the samples.

Whenever you see two elements run together, you should multiply. Thus, *XY* means to multiply the values *X* and *Y* together. $n\Sigma XY$ means to multiple the value n by the value of ΣXY.

The $\sqrt{\ }$ symbol is the square-root operator. When you see $\sqrt{n}$, you need to find the value that, when multiplied by itself, yields *n*.

For example, $\sqrt{9}$ is 3, because 3×3 = 9. Most calculators allow you to easily compute square roots. Another way to express 3×3, by the way, is to write it as 32. The small, raised 2 indicates that you should multiply the value by itself.

Finally, in our notation, we use a horizontal bar over a value to indicate the mean of a sample. For example: indicates that you should compute the mean of the values for all cases in the sample labeled *X*.

SUMMATION NOTATION

The Greek letter Σ (a capital sigma) is used to designate summation. For example, suppose an experimenter measured the performance of four subjects on a memory task. Subject 1's score will be referred to as $X_{1,}$ Subject 2's as $X_{2,}$ and so on.

The scores are shown below:

Subject		Score
1	X_1	7
2	X_2	6
3	X_3	5
4	X_4	8

The way to use the summation sign to indicate the sum of all four X's is:

$$\sum_{i=1}^{4} X_i$$

This notation is read as follows: Sum the values of X from X_1 through X_4. The index i (shown just under the Σ sign) indicates which values of X are to be summed. The index i takes on values beginning with the value to the right of the "=" sign (1 in this case) and continues sequentially until it reaches the value above the Σ sign (4 in this case). Therefore i takes on the values 1, 2, 3, and 4 and the values of X_1, X_2, X_3, and X_4 are summed (7 + 6 + 5 + 8 = 26). In order to make formulas more general, variables can be used with the summation notation.

For example,

$$\sum_{i=1}^{N} X_i$$

means to sum up values of X from 1 to N where N can be any number but usually indicates the sample size. Often an abbreviated form of the summation notation is used. For example, ΣX means to sum all the values of X. When only a subset of the values of X are to be summed then the full version is required. Thus, the sum of all elements of X except the first and the last (the N'th) would be indicated as:

$$\sum_{i=2}^{N-1} X_i$$

which would be read as the sum of X with i going from 2 to

N-1. Some formulas require that each number be squared before the numbers are summed.

This is indicated by:

$$\sum_{i=2}^{N} X_i^2$$

and is equal to $7^2 + 6^2 + 5^2 + 8^2 = 174$. The abbreviated version is simply: ΣX^2. It is very important to note that it makes a big difference whether the numbers are squared first and then summed or summed first and then squared. The symbol $(\Sigma X)^2$ indicates that the numbers should be summed first and then squared. For the present example, this equals: $(7 + 6 + 5 + 8)^2 = 26^2 = 676$. This, of course, is quite different from 174. Sometimes a formula requires that the sum of cross products be computed. For instance, if 3 subjects were each tested twice, they might each have a score on X and on Y.

Subject	X	Y
1	2	3
2	1	6
3	4	5

The sum of cross products (2 x 3) + (1 x 6) + (4 x 5) = 32 can be represented in summation notation simply as: ΣXY.

Basic Theorems: The following data will be used to illustrate the theorems:

X	Y
3	8
2	3
4	1

$\Sigma(X + Y) = \Sigma X + \Sigma Y$
$\Sigma(X + Y) = 11 + 5 + 5 = 21$
$\Sigma X = 3 + 2 + 4 = 9$
$\Sigma Y = 8 + 3 + 1 = 12$
$\Sigma X + \Sigma Y = 9 + 12 = 21$
$\Sigma aX = a\Sigma X$ (a is a constant)
For an example, let a = 2.

$\Sigma aX = (2)(3) + (2)(2) + (2)(4) = 18$

$a\ \Sigma X = (2)(9) = 18$

$\Sigma(X\text{-}M)^2 = \Sigma X^2\text{-}(\Sigma X)^2/N$

(N is the number of numbers, 3 in this case, and M is the mean which is also equal to 3 in this case.

$\Sigma(X\text{-}M)^2 = (3\text{-}3)^2 + (2\text{-}3)^2 + (4\text{-}3)^2 = 2$

$\Sigma X^2 = 3^2 + 2^2 + 4^2 = 29$

$(\Sigma X)^2/N = 9^2/3 = 27$

$\Sigma X^2\text{-}(\Sigma X)^2/N = 29\text{-}27 = 2$

MEAN

A *mean* is a type of average, often referred to as an *arithmetic average*. Given a sample set of values, you calculate the mean by summing all the values and dividing the result by the sample size. Consider the following set of height values for seven kids in a hypothetical fifth-grade class:

100 cm

100 cm

100 cm

110 cm

115 cm

120 cm

180 cm

Given these values, you compute the mean height by summing the values and dividing by the sample size of seven.

For example:

100 + 100 + 100 + 110 + 115 + 120 + 180 = 825 cm

825 cm/7 = 117.86 cm

In this case, the mean height is 117.86 cm. As you can see, the mean height may not correspond to any student's actual height.

Arithmetic Mean

The arithmetic mean is probably the most commonly used statistic in all social science research. The mean is simply the arithmetic average of a distribution of scores, and researchers like it because it provides a single, simple number that gives a rough summary of the distribution. It is important to remember that although the mean provides a useful piece of information, it does not tell you anything about how spread out the scores are (*i.e.,* variance) or how many scores in the distribution are close to the mean.

It is possible for a distribution to have very few scores at or near the mean.

The arithmetic mean is what is commonly called the average: When the word "mean" is used without a modifier, it can be assumed that it refers to the arithmetic mean. The mean is the sum of all the scores divided by the number of scores. The formula in summation notation is:

$$\Sigma = \Sigma X/N$$

where μ is the population mean and N is the number of scores.

If the scores are from a sample, then the symbol M refers to the mean and N refers to the sample size. The formula for M is the same as the formula for μ.

$$M = \Sigma X/N$$

The mean is a good measure of central tendency for roughly symmetric distributions but can be misleading in skewed distributions since it can be greatly influenced by scores in the tail.

Therefore, other statistics such as the median may be more informative for distributions such as reaction time or family income that are frequently very skewed

The sum of squared deviations of scores from their mean is lower than their squared deviations from any other number. For normal distributions, the mean is the most efficient and therefore the least subject to sample fluctuations of all measures of central tendency.

The formal definition of the arithmetic mean is $\mu = E[X]$ where μ is the population mean of the variable X and E[X] is the expected value of X.

MEDIAN

A *median* is yet another type of statistic that you could loosely say refers to the middle value in a sample set. To be more precise, the median is the value in a sample set that has the same number of cases below it as above it.

Consider again the following height data, which is repeated from the earlier section on means:

100 cm
100 cm
100 cm
110 cm
115 cm
120 cm
180 cm

There are seven cases total. Look at the 110 cm value. There are three cases with values above 110 cm and three cases with values below 110 cm. The median height, therefore, is 110 cm.

Computing a median with an odd number of cases is relatively straightforward. Having an even number of cases, though, makes the task of computing a median more complex. With an even number of cases, you simply can't pick just one value that has an equal number of cases above and below it.

The following sample amply illustrates this point:

100 cm
100 cm
100 cm
110 cm
115 cm
120 cm
125 cm
180 cm

There is no "middle" value in this sample. 110 cm won't work because there are four cases having higher values and only three having lower values. 115 cm won't work for the opposite reason—there are three cases having values above and four cases having values below.

In a situation like this, there are two approaches you can take. One solution is to take the lower of the two middle values as the median value. In this case, it would be 110 cm, and it would be referred to as a *statistical median*. Another solution is to take the mean of the two middle values. In this case, it would be 112.5 cm [(115 cm + 110 cm)/2]. A median value computed by taking the mean of the two middle cases is often referred to as a *financial median.*

It is actually possible to find an example where there is a median with an even number of cases. Such an example is: 1,2,3,3,4,5. There is an even number of cases in the sample; however, 3 has an equal number of cases above and below it, so that is the median. This example is exceptional, though, because the two middle cases have the same value. You can't depend on that occurring all the time.

The median is the score in the distribution that marks the 50th percentile. That is, 50% per cent of the scores in the distribution fall above the median and 50% fall below it. Researchers often use the median when they want to divide their distribution scores into two equal groups (called a median split). The median is also a useful statistic to examine when the scores in a distribution are skewed or when there are a few extreme scores at the high end or the low end of the distribution.

The median is the middle of a distribution: half the scores are above the median and half are below the median. The median is less sensitive to extreme scores than the mean and this makes it a better measure than the mean for highly skewed distributions. The median income is usually more informative than the mean income, for example. The sum of the absolute deviations of each number from the median is lower than is the sum of absolute deviations from any other number.

The mean, median, and mode are equal in symmetric distributions. The mean is typically higher than the median in positively skewed distributions and lower than the median in negatively skewed distributions, although this may not be the case in bimodal distributions.

Computation of Median

When there is an odd number of numbers, the median is simply the middle number. For example, the median of 2, 4, and 7 is 4. When there is an even number of numbers, the median is the mean of the two middle numbers. Thus, the median of the numbers 2, 4, 7, 12 is (4 + 7)/2 = 5.5.

MODE

A *mode* is another type of statistic that refers to the most frequently occurring value in a sample set. Look again at the numbers in the previous section on means.

You'll see that the number of occurrences is as follows:

100 cm three
110 cm one
115 cm one
120 cm one
180 cm one

It's obvious that the most frequently occurring value is 100 cm. That value represents the mode for this sample set. The mode is the least used of the measures of central tendency because it provides the least amount of information. The mode simply indicates which score in the distribution occurs most often, or has the highest frequency.

The mode is the most frequently occurring score in a distribution and is used as a measure of central tendency. The advantage of the mode as a measure of central tendency is that its meaning is obvious. Further, it is the only measure of central tendency that can be used with nominal data.

The mode is greatly subject to sample fluctuations and is therefore not recommended to be used as the only measure of central tendency. A further disadvantage of the mode is that many distributions have more than one mode. These distributions are called "multi modal." In a normal distribution, the mean, median, and mode are identical.

STANDARD ERROR

A *standard error* is a measurement of a statistical projection's accuracy based on the sample size used to obtain that projection. Suppose that you were trying to calculate the average height of students in the fifth grade. You could randomly select seven children and calculate their average height.

You could then use that average height as an approximation of the average height for the entire population of fifth graders. If you select a different set of seven children, your results will probably be slightly different. If you increase your sample size, the variation between samples should decrease, and your approximations should become more accurate. In general, the greater your sample size, the more accurate your approximation will be. You can use the sample size and the standard deviation to estimate the error of approximation in a statistical projection by computing the *standard error of the mean.*

The formula to use is:

$$\frac{s}{\sqrt{n}}$$

where:

s Is the standard deviation in the sample.

n Is the number of cases in the sample.

We can use the data shown in the previous section on standard deviation in an example here showing how to compute the standard error. Recall that the sample size was seven elements and that the standard deviation was 28.56.

The standard error, then, is calculated as follows:

$$\frac{28.56}{\sqrt{7}} = \frac{28.56}{2.65} = 10.7$$

Double the sample size to 14 elements, and you have the following:

$$\frac{28.56}{\sqrt{14}} = \frac{28.56}{3.74} = 7.6$$

A standard-error value by itself doesn't mean much. It only has meaning relative to other standard-error values computed for different sample sizes for the same population. In the examples shown here, the standard error is 10.78 for a sample size of 7 and 7.64 for a sample size of 14. A sample size of 14, then, yields a more accurate approximation than a sample size of 7.

POPULATIONS AND SAMPLES

There are two different symbols used for the mean, X and μ. Two different symbols are needed because it is important to distinguish between a statistic that applies to a sample and a parameter that applies to a population. A population is an individual or group that represents all the members of a certain group or category of interest.

A sample is a subset drawn from the larger population. For example, let us suppose that wanted to know the average income of the current full-time, tenured faculty at Harvard. There are two ways that we can find this average. First, get a list of every full-time, tenured faculty member at Harvard and find out the annual income of each member on this list.

Because this list contains every member of the group is interested in, it can be considered a population. If we have to collect these data and calculate the mean, we would have generated a parameter, because a parameter is a value generated from, or applied to, a population. The symbol used to represent the population mean is u. Another way to generate the mean income of the tenured faculty at Harvard would be to randomly select a sub-set of faculty names from our list and calculate the average income of this sub-set.

The sub-set is known as a sample (in this case it is a random sample), and the mean that is generated from this sample is a type of statistic. Statistics are values derived from sample data, whereas parameters are values that are either derived from, or applied to, population data. It is important to note that all samples are representative of some population and that all sample statistics can be used as estimates of population parameters. In the case of the mean, the sample statistic is represented with the symbol. The distinction between sample statistics and population parameters appears again in later chapters.

MEASURES OF CENTRAL TENDENCY

The calculations for each measure of central tendency are mercifully straightforward. With the aid of a calculator or statistics software programme, you will probably never need to calculate any of these statistics. But for the sake of knowledge and in the event you find yourself without a calculator and in need of these statistics, here is the information you will need. Because the mean is an average, calculating the mean involves adding, or summing, all of the scores in a distribution and dividing by the number of scores. So, if you have 10 scores in a distribution, you would add all of the scores together to find the sum and then divide the sum by 10, which is the number of scores in the distribution.

$$\overline{X}, \mu = \frac{\Sigma X}{n, N}$$

where $\overline{X}$ is the sample mean,

μ is the population mean
Σ means "the sum of"
X is an individual score in the distribution
n is the number of scores in the sample
N is the number of scores in the population

The calculation of the median for a simple distribution of scores is even simpler than the calculation of the mean. To find the median of a distribution, you need to first arrange all of the scores in the distribution in order, from smallest to largest.

Once this is done, you simply need to find the middle score in the distribution. If there are an odd number of scores in the distribution, there will be a single score that marks the middle of the distribution. For example, if there are 11 scores in the distribution arranged in descending order from smallest to largest, the 6th score will be the median because there will be 5 scores below it and 5 scores above it.

However, if there are an even number of scores in the distribution, there is no single middle score. In this case, the median is the average of the two scores in the middle of the distribution (as long as the scores are arranged in order, from largest to smallest).

For example, if there are 10 scores in a distribution, to find the median you will need to find the average of the 5th and 6th scores. To find this average, you simply add the two scores together and divide by two. To find the mode, there is no need to calculate anything. The mode is simply the category in the distribution that has the highest number of scores, or the highest frequency. For example, suppose you have the following distribution of IQ test scores from 10 students: 86 90 95 100 100 100 110 110 115 120. In this distribution, the score that occurs most frequently is 100, making it the mode of the distribution. If a distribution has more than one category with the most common score, the distribution has multiple modes and is called multimodal.

One common example of a multimodal distribution is the bimodal distribution. Researchers often get bimodal distributions when they ask people to respond to controversial questions that tend to polarize the public. For example, if we were to ask a sample of 100 people how they feel about capital punishment, we might get the results presented.

On the following scale, please indicate how you feel about capital punishment.

1———2———3———4———5

Strongly opposed — Strongly in Favour

In this example, because most people either strongly opposed or strongly supported capital punishment, we ended up with a bimodal distribution of scores.

EXAMPLE

The Mean, Median, and Mode of a Skewed Distribution

When scores in a distribution are normally distributed, the mean, median, and mode are all at the same point: the centre of the distribution. In the messy world of social science, however, the scores from a sample on a given variable are often not normally distributed. When the scores in a distribution tend to bunch up at one end of the distribution and there are a few scores at the other end, the distribution is said to be skewed. When working with a skewed distribution, the mean, median, and mode are usually all at different points.

It is important to note that the procedures used to calculate a mean, median, and mode are the same whether you are dealing with a skewed or a normal distribution. All that changes are where these three measures of central tendency are in relation to each other. To illustrate, this a fictional distribution of scores is created based on a sample size of 30. Suppose that we were to ask a sample of 30 randomly selected fifth graders whether they think it is important to do well in school. Suppose, further it is asked them to rate how important they think it is to do well in school using a 5-point scale, with 1= "not at all important" and 5 = "very important" Because most fifth graders tend to believe it is very important to do well in school, most of the scores in this distribution are at the high end of the scale, with a few scores at the low end. We have arranged our fictitious scores in order from smallest to largest and get the following distribution:

1	1	1	2	2	2	3	3	3	3
4	4	4	4	4	4	4	4	5	5
5	5	5	5	5	5	5	5	5	5

As you can see, there are only a few scores near the low end of the distribution (1's and 2's) and more at the high end of the distribution (4's and 5's). To get a clear picture of what this skewed distribution looks like, the graph in Figure is created.

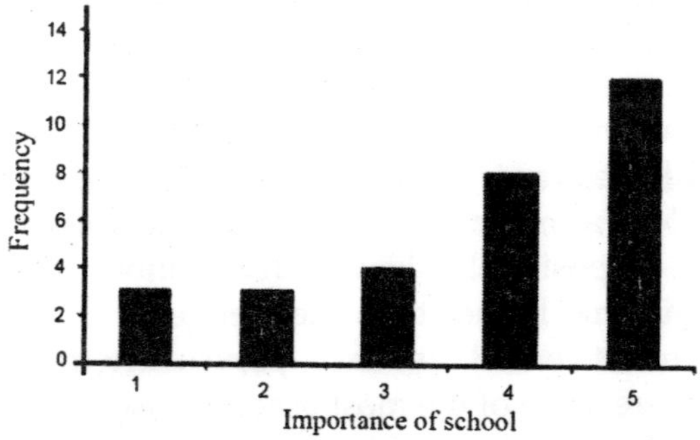

A skewed distribution. This graph provides a picture of what some skewed distributions look like. Notice how most of the scores are clustered at the higher end of the distribution and there are a few scores creating a tail towards the lower end. This is known as a negatively skewed distribution, because the tail goes towards the lower end. If the tail of the distribution were pulled out towards the higher end, this would have been a positively skewed distribution. A quick glance at the scores in the distribution, or at the graph, reveals that the mode is 5 because there were more scores of 5 than any other number in the distribution.

To calculate the mean, we simply apply the formula mentioned earlier. That is, we add up all of the scores (EX) and then divide this sum by the number of scores in the distribution (n). This gives us a fraction of 113/30, which reduces to 3.7666. When we round to the second place after the decimal, we end up with a mean of 3.77.

To find the median of this distribution, we arrange the scores in order from smallest to largest and find the middle score. In this distribution, there are 30 scores, so there will be 2 in the middle. When arranged in order, the 2 scores in the middle (the 15th and 16th scores) are both 4. When we add these two scores together and divide by 2, we end up with 4, making our median 4.

As mentioned earlier, the mean of a distribution can be affected by scores that are unusually large or small for a distribution, sometimes called outliers, whereas the median is not affected by such scores. In the case of a skewed distribution, the mean is usually pulled in the direction of the tail, because the tail is where the outliers are. In a negatively skewed distribution, such as the one presented previously, we would expect the mean to be smaller than the median, because the mean is pulled towards the tail whereas the median is not. In our example, the mean (3.77) is somewhat lower than the median (4). In positively skewed distributions, the mean is usually somewhat higher than the median.

Measures of central tendency, particularly the mean and the median, are some of the most used and useful statistics for researchers. They each provide important information about an entire distribution of scores in a single number. For example, we know that the average height of a man in the United States is five feet nine inches tall.

This single number is used to summarize information about millions of men in this country. But for the same reason that the mean and median are useful, they can often be dangerous if we forget that a statistic such as the mean *ignores* a lot of information about a distribution, including the great amount of variety that exists in many distributions. Without considering the variety as well as the average, it becomes easy to make sweeping generalizations, or stereotypes, based on the mean.

7

Distribution

NORMAL DISTRIBUTION

The normal distribution is a concept with which most people have some familiarity, although they often have never heard of the term. A more familiar name for the normal distribution is the bell curve, because a normal distribution forms the shape of a bell. The normal distribution is extremely important to statistics, and has some specific characteristics that make it so useful. In this chapter it is derscribed what a normal distribution is and why it is so important to researchers. Then some of the features of the normal distribution, and of sampling is discussed, in more depth.

Normal distributions are a family of distributions that have the same general shape. They are symmetric with scores more concentrated in the middle than in the tails. Normal distributions are sometimes described as bell shaped.

Examples of normal distributions are shown below. Notice that they differ in how spread out they are. The area under each curve is the same. The height of a normal distribution can be specified mathematically in terms of two parameters: the mean (μ) and the standard deviation (σ).

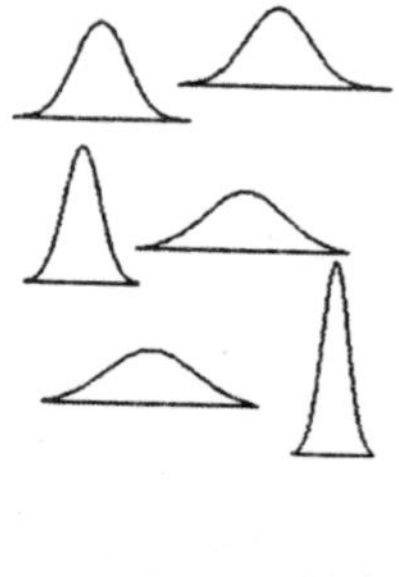

CHARACTERISTICS OF THE NORMAL DISTRIBUTION

If you take a look at the normal distribution shape presented in Figure, you may notice that the normal distribution has three fundamental characteristics. First, it is symmetrical, meaning that the upper half and the lower half of the distribution are mirror images of each other. Second, the mean, median, and mode are all in the same place, in the centre of the distribution (*i.e.*, the top of the bell curve). Because of this second feature, the normal distribution is highest in the middle, it is unimodal, and it curves downward towards the top and bottom of the distribution. Finally, the normal distribution is asymptotic, meaning that the upper and lower tails of the distribution never actually touch the baseline, also known as the *X* axis.

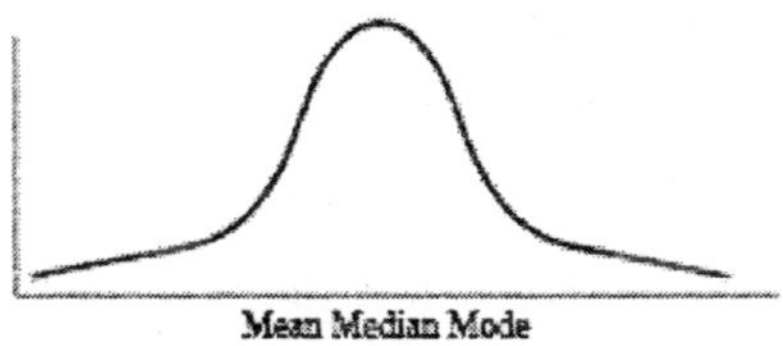

The normal distribution.

FORMULA FOR HEIGHT OF A NORMAL DISTRIBUTION

The height (ordinate) of a normal curve is defined as:

$$\frac{1}{\sqrt{2\pi\sigma^2}} e^{\frac{-(X-\mu)^2}{2\sigma^2}}$$

where μ is the mean and σ is the standard deviation, ð is the constant 3.14159, and e is the base of natural logarithms and is equal to 2.718282. *x* can take on any value from -infinity to +infinity. *f*(*x*) is very close to 0 if x is more than three standard deviations from the mean (less than -3 or greater than +3).

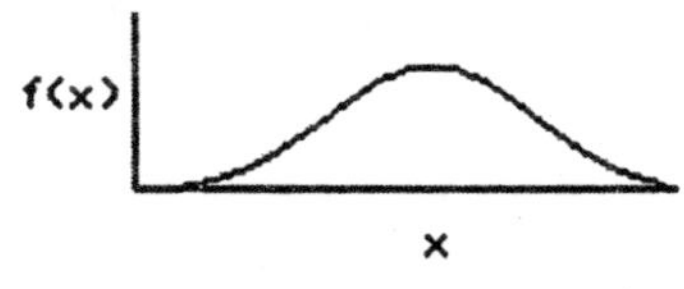

STANDARD NORMAL DISTRIBUTION

The standard normal distribution is a normal distribution with a mean of 0 and a standard deviation of 1. Normal distributions can be transformed to standard normal distributions by the formula:

$$z = \frac{X - \mu}{\sigma}$$

where X is a score from the original normal distribution, μ is the mean of the original normal distribution, and σ is the standard deviation of original normal distribution. The standard normal distribution is sometimes called the z distribution.

A z score always reflects the number of standard deviations above or below the mean a particular score is. For instance, if a person scored a 70 on a test with a mean of 50 and a standard deviation of 10, then they scored 2 standard deviations above the mean.

Converting the test scores to z scores, an X of 70 would be:

$$z = \frac{70 - 50}{10} = 2$$

So, a z score of 2 means the original score was 2 standard deviations above the mean. Note that the z distribution will only be a normal distribution if the original distribution (X) is normal.

Applying the formula

$$z = \frac{X - \mu}{\sigma}$$

will always produce a transformed distribution with a mean of zero and a standard deviation of one. However, the shape of the distribution will not be affected by the transformation.

If X is not normal then the transformed distribution will not be normal either. One important use of the standard normal distribution is for converting between scores from a normal distribution and percentile ranks.

Areas under portions of the standard normal distribution are shown below. About 0.68 (0.34 + 0.34) of the distribution is between -1 and 1 while about 0.96 of the distribution is between -2 and 2.

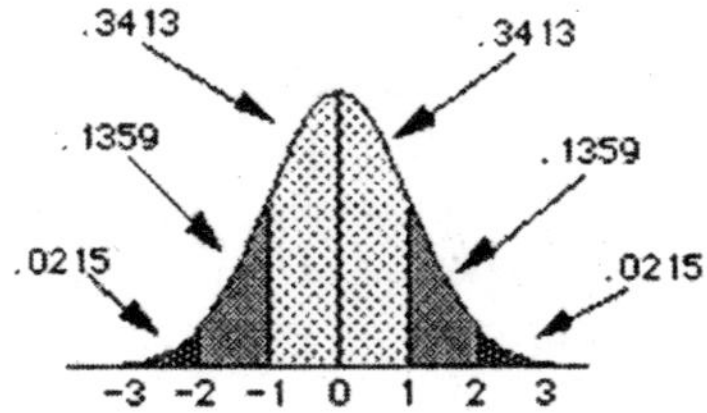

IMPORTANCE OF NORMAL DISTRIBUTION

These three characteristics of the normal distribution are each critical in statistics because they allow us to make good use of probability statistics. When researchers collect data from a sample, sometimes all they want to know about are characteristics of the sample. For example, if we wanted to examine the eating habits of 100 first-year college students, we would just select 100 students, ask them what they eat, and summarize our data.

These data might give me statistics such as the average number of calories consumed each day by the 100 students in our sample, the most commonly eaten foods, the variety of foods eaten, and so on. All of these statistics simply *describe* characteristics of our sample, and are therefore called descriptive statistics. Descriptive statistics generally are used only to describe a specific sample. When all we care about is describing a specific sample, it does not matter whether the scores from the sample are normally distributed or not.

Many times, however, researchers want to do more than simply describe a sample. Sometimes, they want to know what the exact probability is of something occurring in their sample just due to chance. For example, if the average student in our sample consumes 2,000 calories a day, what are the chances, or probability, of having a student in the sample who consumes 5,000 calories a day? In addition, researchers often want to be able to make inferences about the population based on the data they collect from their sample. To determine whether some phenomenon observed in a sample represents an actual phenomenon in the population from which the sample was drawn, inferential statistics are used.

For example, if we find that the men in our sample eat an average of 200 more calories a day than the women, we may want to know the likelihood, or probability, that this gender difference that we have in our sample represents a real difference between the larger populations of men and women in their first year of college. To calculate these probabilities, we need to rely on the normal distribution, because the characteristics of the normal distribution allow statisticians to generate exact probability statistics. In the next section, we will briefly explain how this works.

It is important to note that the normal distribution is what is known in statistics as a theoretical distribution. That is, one rarely, if ever, gets a distribution of scores from a sample that forms an exact, normal distribution. Rather, what you get when you collect data is a distribution of scores that may or may not approach a normal, bell-shaped curve.

Because the theoretical normal distribution is what statisticians used to develop the probabilities discussed earlier, a distribution of scores that is not normal may be at odds with these probabilities. Therefore, there are a number of statistics that begin with the assumption that scores are normally distributed. When this assumption is violated (*i.e.*, when the scores in a distribution are not normally distributed), there can be dire consequences.

The most obvious consequence of violating the assumption of a normal distribution is that the probabilities associated with a normal distribution are not valid. For example, if you have a normal distribution of scores on some variable (*e.g.*, IQ test scores of adults in the United States), you can use the probabilities based on the normal distribution to determine exactly what percentage of the scores in the distribution will be 120 or higher on the IQ test. But suppose the scores in our distribution do not form a normal distribution. Suppose, for some reason, we have an unusually large number of high scores (*e.g.*, over 120) and an unusually small number of low scores (*e.g.*, below 90) in our distribution.

If this were the case, when we use probability estimates based on the normal distribution, we would underestimate the actual number of high scores in our distribution and overestimate the actual number of low scores in our distribution.

SAMPLING ISSUES

The problem of violating the assumption of normality becomes most problematic when our sample is not an adequate representation of our population. In most social science research, we do not have access to an entire population. Therefore, we usually need to rely on a sample for our data, and we must hope that our sample represents our population. ofcourse, to select a sample that represents our population, we must begin by very carefully defining our population.

For example, if we conduct a study in California to examine students' knowledge of world geography, and we select a sample of 500 fifth-grade students from two elementary schools in Los Angeles, what is our population? Do they represent all fifth graders in those two schools? Or all fifth graders in Los Angeles? Or all fifth graders in public schools in Los Angeles? Or in the state? Or in the country?

It turns out that this is a very difficult question to answer. There are all kinds of reasons why a sample may not be a good representation of the larger population that we think it is supposed to represent. Returning to our sample of 500 fifth graders from Los Angeles, suppose that we are trying to make the claim that this sample represents all fifth grade students in Los Angeles public schools. Why might this not be true? For one thing, the two schools from which these students were selected may be different from the typical Los Angeles public elementary school. They may have a larger number of wealthy students than the average school, for example. Or they may spend an unusually small amount of time discussing geography in their fifth grade classrooms. Or they may have an inordinate number of diplomats' children.

On the other hand, these students in our sample may be an excellent representation of the larger population of fifth graders in Los Angeles public schools. Even if they may differ in their wealth or number of students in the classroom or what they eat for breakfast, as long as they do not differ in some *systematic* way *on the variable of interest* (*i.e.*, geography knowledge), there is no reason to believe that this sample does not represent this population.

So the problem of defining our target population is somewhat tricky. Once we have done this, the task becomes one of selecting a sample that has the best chance of representing our population, if possible. One method researchers have adopted is called random sampling. Random sampling occurs if members of the sample to be examined in the study are selected *at random* from the population. We should note that the term *at random* has a somewhat different meaning in statistics than in everyday language. It does not mean haphazardly.

Rather, it means that every member of the population has an equal chance of being selected for inclusion in the sample. Although this may sound easy, in many cases it is quite difficult. For example, if we define our population as all adults in the United States, how could we make sure that every adult has an equal chance of being selected for our sample? We cannot simply select names out of a phonebook, because not everyone has a phone and those who do not own a phone are more likely to be poor than those who do. We cannot simply select names off of voter registration lists because not everyone is registered to vote, and those who are registered are generally better educated than those who are not. As we can see, it can be difficult to get a truly random sample from a population.

A second method of sampling sometimes used by researchers is called representative sampling. In this method, the researcher selects one or more traits to use in selecting a sample from the population. For example, if we are interested in looking at issues of gender and ethnicity, we may try to select a sample that represents the national population (*i.e.*, a sample that is 52% female, 11% African American, etc.). Unfortunately, this can be a costly and time-consuming way of selecting a sample, so many researchers decide against representative sampling.

A third, quite common method of selecting a sample is called convenience sampling. In this method, a sample is selected because there was relatively easy access to the members of the sample. For example, if we were to conduct a study of student achievement, we may select a school that is fairly close to where we live or work so that we have less difficulty selecting the data. In addition, human beings have the option of declining participation in a study, so we must select a school that agrees to participate in our study.

Once we find a relatively close school that agrees to participate in our study, we may be working with a group of wealthy Lithuanian immigrants who are quite different from the larger population of students in our city, state, and country. With convenience sampling, it is often difficult to define the target population before identifying the sample. Rather, a sample is selected and its characteristics are noted (*e.g.*, race, socioeconomic status, any unique or special characteristics) so that the population that the sample represents can later be inferred.

Sampling Method and the Normal Distribution

The relationship between the normal distribution and sampling methods is as follows. The probabilities generated from the normal distribution depend on (a) the shape of the distribution, and (b) the idea that the sample is not somehow systematically different from the population.

If we select a sample randomly from a population, we know that this sample may not look the same as another sample of equal size selected randomly from the same population. But any differences between our sample and other random samples of the same size selected from the same population would differ from each other randomly, not systematically. In other words, our sampling method was not biased such that we would continually select a sample from one end of our population (*e.g.*, the more wealthy, the better educated, the higher achieving) if we continued using the same method for selecting our sample. Contrast this with a convenience sampling method. If we only select schools that are near our home or work, we will continually select schools with similar characteristics. For example, if we live in the Bible Belt, our sample will probably be biased in that our sample will more likely hold fundamentalist religious beliefs than the larger population of schoolchildren. Now if this characteristic is not related to the variable we are studying (*e.g.*, achievement), then it may not matter that our sample is biased in this way. But if this bias is related to our variable of interest, then we may have a problem.

SKEW AND KURTOSIS

A distribution is skewed if one of its tails is longer than the other. The first distribution shown has a positive skew. This means that it has a long tail in the positive direction. The distribution below it has a negative skew since it has a long tail in the negative direction. Finally, the third distribution is symmetric and has no skew. Distributions with positive skew are sometimes called "skewed to the right" whereas distributions with negative skew are called "skewed to the left."

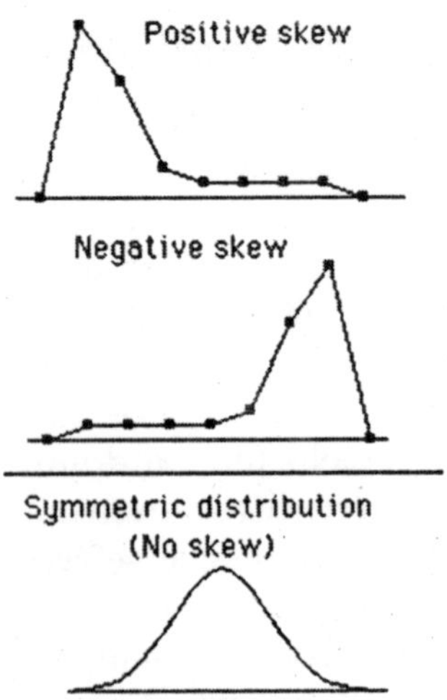

Distributions with positive skew are more common than distributions with negative skews. One example is the distribution of income. Most people make under $40,000 a year, but some make quite a bit more with a small number making many millions of dollars per year. The positive tail therefore extends out quite a long way whereas the negative tail stops at zero. For a more psychological example, a distribution with a positive skew typically results if the time it takes to make a response is measured. The longest response times are usually much longer than typical response times whereas the shortest response times are seldom much less than the typical response time. A histogram of the author's performance on a perceptual motor task in which the goal is to move the mouse to and click on a small target as quickly as possible is shown below. The X axis shows times in milliseconds.

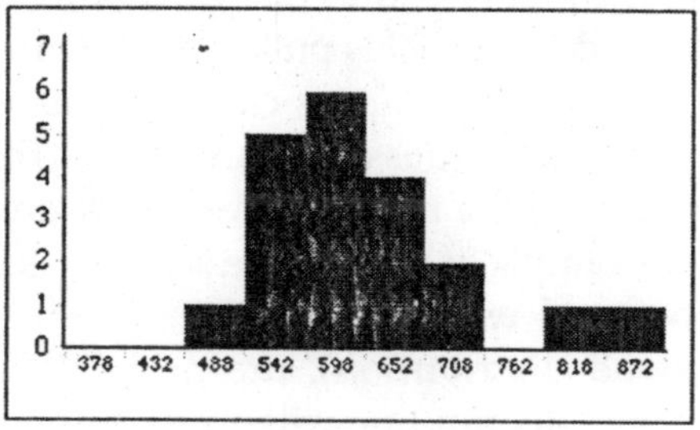

Negatively skewed distributions do occur, however. Consider the following frequency polygon of test grades on a statistics test where most students did very well but a few did poorly. It has a large negative skew.

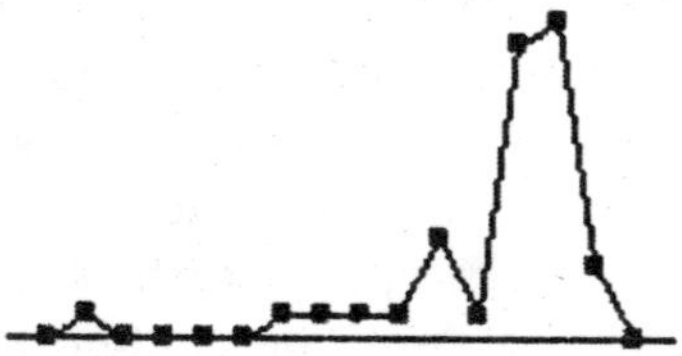

Skew can be calculated as:

$$skew = \frac{\sum (X - \mu)^3}{N\sigma^3}$$

where μ is the mean and σ is the standard deviation. The normal distribution has a skew of 0 since it is a symmetric distribution. As a general rule, the mean is larger than the median in positively skewed distributions and less than the median in negatively skewed distributions. There are counter examples. For example it is not uncommon for the median to be higher than the mean in a postively skewed bimodal distribution or with discrete distributions.

Two characteristics used to describe a distribution of scores are skew and kurtosis. When a sample of scores is not normally distributed (*i.e.,* the bell shape), there are a variety of shapes it can assume. One way a distribution can deviate from the bell shape is if there is a bunching of scores at one end and a few scores pulling a tail of the distribution out towards the other end. If there are a few scores creating an elongated tail at the higher end of the distribution, it is said to be positively skewed. If the tail is pulled out towards the lower end of the distribution, the shape is called negatively skewed.

As you might have guessed, skewed distributions can distort the accuracy of the probabilities based on the normal distribution. For example, if most of the scores in a distribution occur at the low end with a few scores at the higher end (positively skewed distribution), the probabilities that are based on the normal distribution will underestimate the actual number of scores at the lower end of this skewed distribution and overestimate the number of scores at the higher end of the distribution. In a negatively skewed distribution, the opposite pattern of errors in prediction will occur.

Kurtosis refers the shape of the distribution in terms of height, or flatness. When a distribution has a peak that is higher than that found in a normal, bell-shaped distribution, it is called leptokurtic. When a distribution is flatter than a normal distribution, it is called platykurtic. Because the normal distribution contains a certain percentage of scores in the middle area (*i.e.,* about 68% of the scores fall between 1 standard deviation above and 1 standard deviation below the mean), a distribution that is either patykurtic or letpkurtic will likely have a different percentage of scores near the mean than will a normal distribution. Specifically, a letptokurtic distribution will probably have a greater percentage of scores closer to the mean and fewer in the upper and lower tails of the distribution whereas a platykurtic distribution will have more scores at the ends and fewer in the middle than will a normal distribution.

Normal Distribution Probabilities

To illustrate some of the difficulties that can arise when we try to apply the probabilities that are generated from using the normal distribution to a distribution of scores that is skewed, we present a distribution of sixth-grade students and scores on a measure of self-esteem. In these data, 677 students completed a questionnaire that included four items designed to measure students' overall sense of self-esteem. Examples of these questions include "On the whole, We are satisfied with ourself" and "We feel we have a number of good qualities." Students responded to each of these four questions using a 5-point rating scale with. "1 = not at all true" and "5 = very true." Students' responses on these four items were then averaged, creating a single self-esteem score that ranged from a possible low of 1 to a possible high of 5. The frequency distribution for this self-esteem variable is presented in Figure.

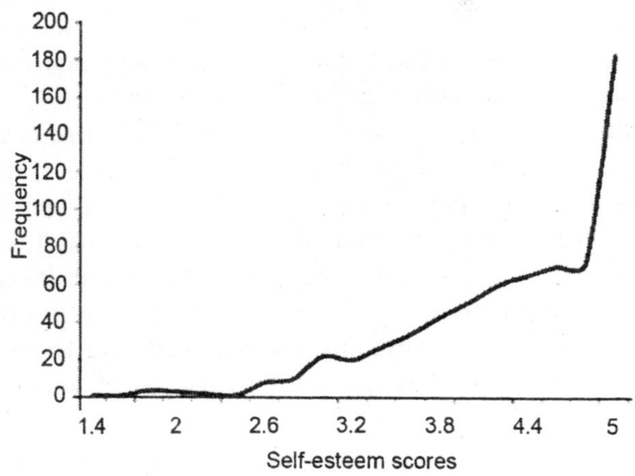

Fig. Frequency Distribution for Self-esteem Scores.

As we can see, the distribution of scores presented in Figure does not form a nice, normal, bell-shaped distribution. Rather, most of the students in this sample scored at the high end of the distribution and a long tail extends out towards the lower end of the scale. This is a classic, negatively skewed distribution of scores. The happy part of this story is that most of the students in this sample appear to feel quite good about themselves. The sad part of the story is that some of the assumptions of the normal distribution are violated by this skewed distribution. Let's take a look at some specifics.

One of the qualities of a normal distribution is that it is symmetrical, with an equal percentage of the scores between the mean and 1 standard deviation below the mean as there are between the mean and 1 standard deviation above the mean. In other words, in a normal distribution, there should be about 34% of the scores within 1 standard deviation above the mean and 34% within 1 standard deviation below the mean.

In our distribution of self-esteem scores presented earlier, the mean is 4.28 and the standard deviation is.72. A full 50% of the distribution falls between the mean and 1 standard deviation above the mean in this group of scores. So, although we might predict that about 16% of our distribution will have scores more than 1 standard deviation above the mean in a normal distribution, in our skewed distribution of self-esteem scores, we can see that there are no students with scores more than 1 standard deviation above the mean.

In the coming we present a more thorough discussion of how to use the normal distribution to calculate standard deviation units and percentile scores in a normal distribution. As this example demonstrates, the probabilities that statisticians have generated using the normal distribution may not apply well to skewed or otherwise non-normal distributions of data. This should not lead you to believe, however, that non-normal distributions of scores are worthless.

In fact, even if you have a non-normal distribution of scores in your sample, these scores can create normal sampling distributions for use in inferential statistics. What is perhaps most important to keep in mind is that a non-normal distribution of scores may be an indication that your sample may differ in important and systematic ways from the population that it is supposed to represent. When making inferences about a population based on a sample, be very careful to define the population precisely and to be aware of any biases you may have introduced by your method of selecting your sample. The normal distribution can be used to generate probabilities about the likelihood of selecting an individual or another sample with certain characteristics (*e.g.*, distance from the mean) from a population.

If your sample is not normal and your method of selecting the sample may be systematically biased to include those with certain characteristics (*e.g.*, higher than average achievers, lower than average income, etc.), then the probabilities of the normal distribution may not apply well to your sample. The theoretical normal distribution is a critical element of statistics primarily because many of the probabilities that are used in inferential statistics are based on the assumption of normal distributions.

As we will see in coming chapters, statisticians use these probabilities to determine the probability of getting certain statistics and to make inferences about the population based on the sample. Even if the data in a sample are not normally distributed, it is possible that the data in the population from which the sample was selected may be normally distributed.

It is described how the normal distribution, through the use of z scores and standardization, is used to determine the probability of obtaining an individual score from a sample that is a certain distance away from the sample mean.

CONCEPT OF STANDARD ERRORS

The concept of standard error is one that many students of statistics find confusing when they first encounter it. In all honesty, there are many students, and many researchers, who never fully grasp the concept. We are convinced that many people have problems with understanding standard errors because they require a bit of a leap into the abstract and because, with the advent of computer programmes, it is possible to lead a long and productive research life without having to think about or analyse a standard error for years at a time.

Therefore, many researchers choose to gloss over this abstract concept. This is a mistake. We hold this opinion because, as a teacher of statistics, We have learned that when one is able to truly understand the concept of standard error, many of our most beloved inferential statistics (*t* tests, ANOVA, regression coefficients, correlations) become easy to understand.

So let me offer this piece of advice: Keep trying to understand the contents of this chapter, and other information you get about standard errors, even if you find it confusing the first or second time you read it. With a little effort and patience, you can understand standard errors and many of the statistics that rely on them.

MEANING OF STANDARD ERROR

There are two answers to this question. First, there is the technical answer, which is the definition of a standard error. A standard error is, in effect, the standard deviation of the sampling distribution of some statistic (*e.g.*, the mean, the difference between two means, the correlation coefficient, etc.). We realise that this makes no sense until you know what a sampling distribution is, and it is explained this in the next section of this chapter. For now, it is recommend that we repeat the definition to yourself 10 times: "The standard error is, in effect, the standard deviation of the sampling distribution of some statistic." The second answer is that the standard error is the denominator in the formulas used to calculate many inferential statistics.

In the following chapters, you will see the standard error as the denominator in many formulas. This is because the standard error is *the measure of how much* random *variation we would expect from samples of equal size drawn from the same population*. Again, look at the preceding sentence, think about it, and rest assured that it is explained in more detail in the next few pages.

Standard Error of the Mean

To begin this more detailed discussion of standard errors, it is introduced that the esoteric component of the concept. This is the section that you may need to read several times to let it sink in. Although there are standard errors for all statistics, we will focus on the standard error of the mean.

When we think of a distribution of scores, we think of a certain number of scores that are plotted in some sort of frequency graph to form a distribution. In these distributions, each case has a score that is part of the distribution. Just as these simple frequency distributions are plotted, or graphed, we can also plot distributions of sample means. Imagine that we wanted to find the average shoe size of adult American women. In this study, the population we are interested in is all adult American women.

But it would be expensive and tedious to measure the shoe size of all adult American women. So we select a sample of 100 women, at random, from our population. At this point, it is very important to realise that our sample of 100 women may or may not look like the typical American woman (in terms of shoe size). When we select a sample at random, it is possible to get a sample that represents an extreme end of the population (*e.g.*, a sample with an unusually large average shoe size). If we were to throw our first sample of women back into the general population and chose another random sample *of the same size* (*i.e.*, 100), it is possible that this second sample may have an average shoe size that is quite different from our first sample.

Once you realise that different random samples of equal size can produce different mean scores on some variable (*e.g.*, different average shoe sizes), the next step in this conceptual puzzle is easy: If we were to take 1,000 different random samples of women, each of 100, and compute the average shoe size of each sample, these 1,000 sample means would form their own distribution. This distribution would be called the sampling distribution of the mean. To illustrate this concept, let's consider an example with a small population (TV = 5). Suppose our population consists of five college students enrolled in a seminar on statistics. Because it is a small seminar, these five students represent the entire population of this seminar.

These students each took the final exam that was scored on a scale from 1 to 10, with lower scores indicating poorer performance on the exam. The scores for each student are presented in Table, arranged in descending order according to how well they did on the exam.

Table. Population of Students' Scores on Final Exam.

Students	Score of Final Exam
Students 1	3
Students 2	6
Students 3	6
Students 4	7
Students 5	9

If we were to select a random sample of two students from this population ($n = 2$), we might get student 2 and student 5. This sample would have a mean of 7.5 because [(6 + 9) + 2 = 7.5]. If we were to put those two students back into the population and randomly select another sample of 2,1 might get Student 4 and Student 5. This sample would have a mean of 8 because [(7 + 9) 2 = 8].

We put those students back into the population and randomly select another sample of 2, such as Students 1 an 3. This sample would have a mean of 4.5.

As we can see, just by virtue of whom is included in each random sample we select from our population, we get different sample means. Now if we were to repeat this process of randomly selecting samples of two students from our population, calculating their mean, and returning the members of the sample to the population (called sampling with replacement), eventually we would get a distribution of sample means that would look something like the distribution presented in figure. As we can see, these means form a distribution. This example illustrates how random samples of a given size selected from a population will produce a distribution of sample means, eventually forming a sampling distribution of the mean.

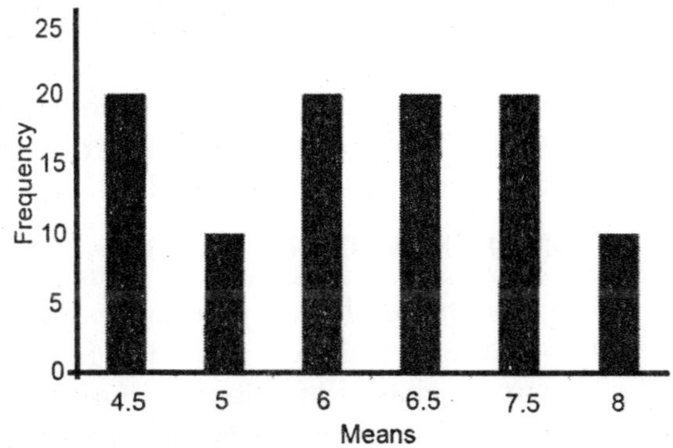

Fig. Sampling Distribution of the Mean.

Just as the other distributions we have discussed have a mean and a standard deviation, this sampling distribution of the mean also has these characteristics. To distinguish a sampling distribution from a simple frequency distribution, the mean and standard deviation of the sampling distribution of the mean have special names. The mean of the sampling distribution of the mean is called the expected value.

It is called the expected value because the mean of the sampling distribution of the means is the same as the population mean. When we select a sample from the population, our best guess is that the mean for the sample will be the same as the mean for the population, so our *expected* mean will be the population mean. The standard deviation of the sampling distribution of the mean is called the standard error. So the standard error is simply the standard deviation of the sampling distribution.

The final step in understanding the concept of standard error of the mean is to understand what this statistic tells us. You will remember that the standard deviation tells us the average difference, or deviation, between an individual score in the distribution and the mean for the distribution.

The standard error of the mean provides essentially the same information, except it refers to the average difference between the expected value (*i.e.*, the population mean) and an individual sample mean. So one way to think about the standard error of the mean is that it tells us how confident we should be that a sample mean represents the actual population mean. Phrased another way, the standard error of the mean provides a measure of how much *error* we can expect when we say that a sample mean represents the mean of the larger population. That is why it is called a standard *error*.

CALCULATION OF STANDARD ERROR

Most of the time, researchers do not draw 1,000 samples of equal size from the population and then figure out the mean and standard deviation of this distribution of sample means. In fact, most of the time, researchers collect data from only a single sample, and then use this sample to make inferences about the population from which the sample was drawn. How can we make inferences about a larger population on the basis of a single sample?

To make such inferences about the population from a single sample, researchers must use what they know about their sample to make educated guesses, or estimates, about the population. demonstrate this concept using the shoe-size example mentioned earlier. Suppose that we have a random sample of 100 women. Now if this sample were truly selected at random (*i.e.*, every adult woman in the United States had an equal chance of being selected), our most logical assumption would be that this sample represents the larger population accurately.

Therefore, we would have to assume that the mean shoe size of our sample (suppose it is 6) is also the mean shoe size of the larger population. Of course, we cannot know if this is true. In fact, as discussed earlier, there is good reason to believe that our sample may not represent our population well. But if the only information we have about U.S. adult women's shoe size comes from our sample of 100 women, our *best guess* about what the larger population of women looks like must be that they are similar to this sample of 100 women. Now we are faced with a critical question: When we guess that the population of women in the United States has an average shoe size of 6 (based on our sample average), how much *error* can we *expect* to have in this estimation? In other words, what is the *standard error?*

To answer this question we must examine two characteristics of our sample. First, how large is our sample? The larger our sample, the less error we should have in our estimate about the population. This makes sense because the larger our sample, the more our sample should look like our population, and the more accurate our estimates of our population will be. If there are 100 million women in the United States and we use a sample of 50 million to predict their average shoe size, we would expect this prediction to be more accurate than a prediction based on a sample of 100 women. Therefore, the larger our sample, the smaller our standard error.

The second characteristic of our sample that we need to examine is the standard deviation. Remember that the standard deviation is a measure of how much variation there is in the scores in our sample. If the scores in our sample are very diverse (*i.e.*, a lot of variation, a large standard deviation), we can assume that the scores in our population are also quite diverse. So, if some women in our sample have size 2 shoes and others have size 14 shoes, we can assume there is also a pretty large variety of shoe sizes in our population. On the other hand, if all of the women in our sample have shoe sizes of either 5, 6, or 7, we can assume that most of the women in the larger population have an equally small variety of shoe sizes.

Of course, these assumptions about the population may not be true, we must rely on them because this is all the information we have. So, the larger the sample standard deviation, the greater the assumed variation of scores in the population, and consequently the larger the standard error of the mean.

$$\sigma_{\bar{x}} = \frac{\sigma}{\sqrt{n}}$$

or

$$s_{\bar{x}} = \frac{s}{\sqrt{n}}$$

Where σ is the standard deviation for the population
s is the sample estimate of the standard deviation
n is the size of the sample

An examination of the formula for calculating the standard error of the mean reveals the central role of the sample standard deviation (or population standard deviation, if known) and the sample size in determining the standard error. The formula is simply the standard deviation of the sample or population divided by the square root of n, the sample size. As with all fractions, as the numerator gets larger, so does the resulting standard error. Similarly, as the size of the denominator decreases, the resulting standard error increases.

Small samples with large standard deviations produce large standard errors, because these characteristics make it more difficult to have confidence that our sample accurately represents our population. In contrast, a large sample with a small standard deviation will produce a small standard error, because such characteristics make it more likely that our sample accurately represents our population.

CENTRAL LIMIT THEOREM

Simply put, the central limit theorem states that as long as you have a reasonably large sample size (*e.g.*, $n > 30$), the sampling distribution of the mean will be normally distributed, even if the distribution of scores in your sample is not. In earlier chapters, it is discussed distributions that were not in the shape of a nice, normal, bell curve. What the central limit theorem proves is that even when you have such a non-normal distribution in your sample, the sampling distribution of the mean will form a nice, normal, bell-shaped distribution as long as you have at least 30 cases in your sample. Even if you have fewer than 30 cases in your sample, the sampling distribution of the mean will probably be near normal if you have at least 10 cases in your sample. Even in our earlier example where we had only two cases per sample, the sampling distribution of the mean had the beginning of a normal shape. Although we do not concern ourselves here with why the central limit theorem works, you need to understand why the veracity of this theorem is so important. A number of statistics rely on probabilities that are generated from normal distributions. For example, we may want to know whether the average IQ test scores of a sample of 50 adults in California is different from the larger population of adults.

If our sample has an average IQ test score of 110, and the national average is 100, we can see that our sample average differs from the population average by 10 points. Is 10 points a meaningful difference or a trivial one? To answer that question, we must be able to discover the probability of getting a difference of 10 points by random chance.

In other words, if we were to select another random sample of 50 adults from California and compute their average IQ test score, what are the odds that they will have an average that is 10 points higher than the national average of 100? To determine this probability, we must have a normal distribution of sample means, or a normal sampling distribution of the mean. The central limit theorem indicates that as long as we have a sample size of at least 30, our sampling distribution of the mean will be normal.

Comparing *z* Scores and *t* Values

To determine the probability of randomly selecting an individual case with a particular score on some variable from a population with a given mean on that variable. We did this by converting the raw score into a z score. Now that we know how to compute a standard error, we can use z scores again to determine the probability of randomly selecting a sample with a particular mean on a variable from a population with a given mean on the same variable. We can also use the family of *t* distributions to generate *t* values to figure out the same types of probabilities. To explain this, we will begin by comparing the normal distribution with the family of *t* distributions. The normal distribution is a theoretical distribution with a bell shape and is based on the idea of population data. We also know that the probabilities associated with z scores are associated with the normal distribution. In addition, we know that a standard deviation derived from sample data is only an *estimate* of the population standard deviation.

Because the formula for calculating the sample standard deviation has n-1 in the denominator, we also know that the smaller the sample, the less precisely the sample standard deviation estimates the population standard deviation. Finally, we know that the standard error formula is based partly on the standard deviation.

When we put all of this information together, we end up with a little bit of the dilemma. If we can use the standard error to generate z scores and probabilities, and these z scores and probabilities are based on the normal distribution, what do we do in those cases where we are using sample data and we have a small sample?

Won't our small sample influence our standard error? And won't this standard error influence our z scores? Will our z scores and probabilities be accurate if we have a small sample? Fortunately, these concerns have already be addressed by brains larger than mine. It turns out that the normal distribution has a close family of relatives: the family of t distributions.

These distributions are very much like the normal distribution, except the shape of t distributions is influenced by sample size. With large samples (*e.g.*, > 120), the shape of the t distribution is virtually identical to the normal distribution.

As sample size decreases, however, the shape of the t distribution becomes flatter in the middle and higher on the ends. In other words, as sample size decreases, there will be fewer cases near the mean and more cases away from the mean, out in the tails of the distribution. Like the normal distribution, t distributions are still symmetrical.

Just as we use the z table (Appendix A) to find probabilities associated with the normal distribution, we use the table of t values (Appendix B) to find probabilities associated with the *t* distributions. Along the left column of Appendix B there are numbers in ascending order.

These are degrees of freedom and they are directly related to sample size. To use this table, you simply calculate a *t* value (using basically the same formula that you use to find a z score) and then, using the appropriate degrees of freedom, figure out where your *t* value falls in Appendix B to determine the probability of finding a *t* value of that size.

Whenever you don't know the population standard deviation and must use an estimate from a sample, it is wise to use the family of *t* distributions. Here is an example to illustrate these ideas.

We used this formula to calculate a z score from a raw score:

$$z = \frac{raw\ score - mean}{s\tan dard\ deviation}$$

or

$$z = \frac{X - \mu}{\sigma}$$

where X = raw score

μ = population mean

σ = standard deviation

The formula for converting a sample mean into a z score is almost identical, except the individual raw score is replaced by the sample mean and the standard deviation is replaced by the standard error. In addition, if we do not know the population standard deviation, the standard deviation estimate from the sample must be used and we are computing a *t* value rather than a z score. These formulas are found in Table.

Now, suppose that we know that the average American man exercises for 60 minutes a week. Suppose, further, that we have a random sample of 144 men and that this sample exercises for an average of 65 minutes per week with a standard deviation of 10 minutes. What is the probability of getting a random sample of this size with a mean of 65 if the actual population mean is 60 *by chance?* To answer this question, we compute a *t* value:

$$t = \frac{65-60}{10/\sqrt{144}}$$

$$t = \frac{5}{.83}$$

$$t = 6.02$$

If we look in Appendix B, using the row with oo degrees of freedom, we can see that the probability of getting a *t* value of this size *by chance* with a sample of this size is less than.001. Notice that if we had calculated a z score rather than a *t* score (*i.e.,* if the *population* standard deviation had been 10), our z value would have been the same (*i.e.,* z = 6.02) and our probability, as found in Appendix A, would have been less than.00003. The normal distribution (associated with z scores) and the *t* distributions are virtually identical when the sample size is larger than 120. Finally, to illustrate the difference between the *t* distributions and the normal distribution, suppose that our sample size had been 25 rather than 144. We would have calculated the *t* value just as we did before, but our standard error would be different (because our sample size is smaller), thereby producing a smaller *t* value:

$$t = \frac{65-60}{10/\sqrt{25}}$$

$$t = \frac{5}{2}$$

$$t = 2.50$$

Now, looking at our table of *t* values with 24 degrees of freedom, we find that the probability of getting a *t* value of this size is just about.02. Notice that if we had our larger sample size of 144, the probability of getting a *t* value of 2.50 would have been closer to.01.

z score and *t* value formulas.

when σ is known

$$z = \frac{sample\,mean - population\,mean}{s\tan dard\,error}$$

$$z = \frac{\overline{X} - \mu}{\sigma \overline{x}}$$

when σ is not known,

$$t = \frac{sample\,mean - population\,mean}{s\tan dard\,error}$$

$$z = \frac{\overline{X} - \mu}{s_{\bar{x}}}$$

Where μ = population mean
σ^2 = standard error
$\overline{X}$ = sample mean
$s_{\bar{x}}$ = sample of the standard error

So when the sample size is large, the normal distribution and the *t* distribution are virtually identical. But as our sample size decreases, the *t* distribution changes and so do the probabilities associated with it. When the population standard deviation is known, the normal distribution can be used. But when the population standard deviation is not know, or the sample size is small, the family of *t* distributions should be used. Now we can turn our attention the how standard errors are used in other statistics.

USE OF STANDARD ERRORS

Some type of standard error is used in every inferential statistic, including those discussed in this book (*t* tests, ANOVA, regression, etc.). In inferential statistics, we want to know whether something that we observe in our sample represents a similar phenomenon in the larger population from which the sample was drawn. For example, if we compare the average weight of a sample 100 men to that of a sample of 100 women and find that, on average, men in our sample weigh 60 pounds more than women in our sample, we may want to know whether we should conclude that, on average, men in the larger population weigh more than women in the population.

Similarly, if we find a correlation of r =.45 between height and weight in our sample of 100 men, we might want to know whether this relationship between height and weight in our sample means there is probably a relationship between these two variables in the larger population of men. To answer these questions, we need to use standard errors.

In many inferential statistics formulas, we need to see whether the phenomenon we observed in our sample(s) is large or small relative to our standard error. A standard error is a measure of the average amount of variance, or difference, we can expect from different samples of the same size selected from a population. So, the question we are asking with many inferential statistics is whether some statistic we see in our sample is big or small compared to the amount of variance (or error) we would expect if we had randomly selected a *different* sample of the same size. This question can be summarized with the following fraction:

$$\frac{\textit{size of sample statistic}}{s\tan dard\,error}$$

As an illustration, let us return to the example comparing the average weight of men and women. We already know that, in our samples, the difference between the average weight of men and women was 60 pounds. The statistic that we am interested in here is the *difference* between the two means (*i.e.*, the average weight of men and the average weight of women). If we were to select two different samples of the same size from the populations of men and women and find the difference in those two sample means, we would probably find a difference that was either larger or smaller than the difference we found in the comparison of the first two samples. If we kept selecting different samples and compared their means, we would eventually get a sampling distribution of the differences between the means, and this sampling distribution would have a standard error. Suppose that the standard error of this sampling distribution was 10. Let's plug that standard error into our fraction formula presented earlier:

$$\frac{\textit{sample statistic}}{s\tan dard\,error} = \frac{60}{60}$$

The difference between our two sample means is six times larger than the difference we would expect to find just due to random sampling error. This suggests that the difference between our two sample means is probably not due to chance. Using a table of probabilities based on the *t* distribution, we can calculate the exact probability of getting a ratio this large (*i.e.*, 60:10, or 6:1).

So, to summarize, the standard error is often used in inferential statistics to see whether our sample statistic is larger or smaller than the average differences in the statistic we would expect to occur by chance due to differences between samples. We now discuss some examples to demonstrate the effect of sample size and the standard deviation on the size of the standard error of the mean.

Effect of Sample Size and Standard Deviation on Standard Error

To illustrate the effect that sample size and standard deviation have on the size of the standard error of the mean, let's take a look at a variable from a set of data we collected a few years ago. The purpose of the study was to examine students' motivational beliefs about standardized achievement tests. We examined whether students thought it was important to do well on the standardized test they were about to take in school, whether they had anxiety about the test, whether they expected to do well on the test, whether they thought of themselves as good test takers, and so on.

One of the goals of the study was to compare the motivational beliefs of elementary school students with those of middle school students.

The sample for the study included 137 fifth graders in elementary school with 536 seventh and eighth graders in middle school. Suppose we wanted to know the standard error of the mean on the variable "We expect to do well on the test" for each of the two groups in the study, the elementary school students and the middle school students. To calculate these standard errors, we would need to know the standard deviation for each group on our variable and the sample size for each group. These statistics are presented in Table. Standard Deviation and Sample Action.

	Standard Dev.	Sample Size	Standard Dev.	Sample Size
Expect to do with	1.34	1.37	1.46	5.36

A quick glance at the standard deviations for each group reveals that they are very similar (s = 1.38 for the elementary school sample, s = 1.46 for the middle school sample). However, because there is quite a large difference in the size of the two samples, we should expect somewhat different standard errors of the mean for each group. Which group do you think will have the larger standard error of the mean?

To find the standard error of the mean, we simply need to divide the standard deviation by the square root of the sample size. For the elementary school sample, we need to divide 1.38 by the square root of 137. The square root of 137 = 11.70. When we divide 1.38 by 11.70 we get.12. So the standard error of the mean for the elementary sample is.12. Following the same procedure for the middle school sample, we find that the standard error of the mean for this group will equal 1.46 divided by the square root of 546. The square root of 546 = 23.37. When we divide 1.46 by 23.37 we get.06.The standard error of the mean for the middle school sample (s.e. =.06) is half the size of the standard error of the mean for the elementary school sample (s.e. =.12). Because the standard deviations were roughly equal for these two groups, virtually all of the difference in their standard errors is attributable to differences in sample size.

To illustrate the effect of the standard deviation on the size of the standard error, let's take a look at a second variable from this study, students' scores on the verbal portion of the standardized achievement tests. Scores on this portion of the test range from a possible low of 0 to a possible high of 100. In the elementary school sample, the standard deviation on this variable was 23.81. The sample size is still 137. To find the standard error of the mean, we must divide 23.81 by the square root of 137 (which we know from our previous example is 11.70. And, 23.81 divided by 11.70 equals 2.04. So the standard error of the mean in this example is 2.04. When we compare this number with the standard error of the mean for the elementary school sample on the "Expect to do well on the test" variable ($s_{\bar{x}}$ =.12), we see that the larger standard deviation for the test score variable created a much larger standard error, even though the sample size remained the same, 137.

As these examples demonstrate, the size of the standard error of the mean depends on the size of the standard deviation and the size of the sample. As sample size increases, the standard error of the mean decreases. As the size of the standard deviation increases, the size of the standard error of the mean increases as well. Remember that the standard error is generally used in the denominator of the formulas statisticians use to calculate inferential statistics. Therefore, smaller standard errors will produce larger statistics (because smaller denominators produce larger overall numbers than larger denominators do when the numerators are equal). Larger statistics are more likely to be statistically significant. Therefore, all else being equal, larger sample sizes are more likely to produce statistically significant results because larger sample sizes produce smaller standard errors.

Standard error is often a difficult concept to grasp the first time it is encountered (or the second or the third). Because it is such a fundamental concept in inferential statistics, however, we encourage you to keep trying to make sense of both the meaning and the usefulness of standard errors.

Standard errors can be used to determine probabilities of sample statistics (such as the mean) in much the same way that we used standard scores to determine probabilities associated with individual scores. Because of the usefulness of standard errors in determining probabilities, standard errors play a critical role in determining whether a statistic is statistically significant. Because standard errors are influenced by sample size, statistical significance will also be influenced by these sample characteristics. The issue of statistical significance, and the effects of sample size on statistical significance, are discussed in more depth.

8

Concept of Correlation

CORRELATION

Researchers compute correlation coefficients when they want to know how two variables are related to each other. For a Pearson product-moment correlation, both of the variables must be measured on an interval or ratio scale, otherwise known as continuous variables. For example, suppose we wanted to know whether there was a relationship between the amount of time students spend studying for an exam and their scores on the exam.

We suspect that the more hours students spend studying, the higher their scores will be on the exam. But we also suspect that there is not a perfect correspondence between time spent studying and test scores. Some students will probably get low scores on the exam even if they study for a long time, simply because they may have a hard time understanding the material.

Indeed, there will probably be a number of students who spend an inordinately long period of time studying for the test precisely *because* they are having trouble understanding the material. On the other hand, there will probably be some students who do very well on the test without spending very much time studying. Despite these "exceptions" to our rule, We will still hypothesize that, *on average,* as the amount of time spent studying increases, so do students' scores on the exam.

There are two fundamental characteristics of correlation coefficients researchers care about. The first of these is the direction of the correlation coefficient. Correlation coefficients can be either positive or negative. A positive correlation indicates that the values on the two variables being analysed move in the same direction.

That is, as scores on one variable go up, scores on the other variable go up as well (on average). Similarly, on average, as scores on one variable go down, scores on the other variable go down. Returning to our earlier example, if there is a positive correlation between the amount of time students spend studying and their test scores, We can tell that, on average, the more time students spent studying, the higher their scores on the test. This is *equivalent* to saying that, on average, the *less* time they spent studying, the *lower* their scores on the test. Both of these represent a *positive* correlation between time spent studying and test scores.

A negative correlation indicates that the values on the two variables being analysed move in *opposite* directions. That is, as scores on one variable go up, scores on the other variable go down, and vice-versa (on average). If there were a negative correlation between the amount of time spent studying and test scores, we would know that, on average, the more time students spend studying for the exam, the *lower* they actually score on the exam.

Similarly, with a negative correlation we would also conclude that, on average, the *less* time students spent studying, the *higher* their scores on the exam. These positive and negative correlations are represented by scattergrams in Figure. Scattergrams are simply graphs that indicate the scores of each case in a sample simultaneously on two variables. For example, in the "Positive Correlation" scattergram in Figure, the first case in the sample studied for 1 hour and got a score of 30 on the exam. The second case studied for 2 hours and scored a 40 on the exam.

The second fundamental characteristic of correlation coefficients is the strength or magnitude of the relationship. Correlation coefficients range in strength from −1.00 to +1.00. A correlation coefficient of.00 indicates that there is no relationship between the two variables being examined.

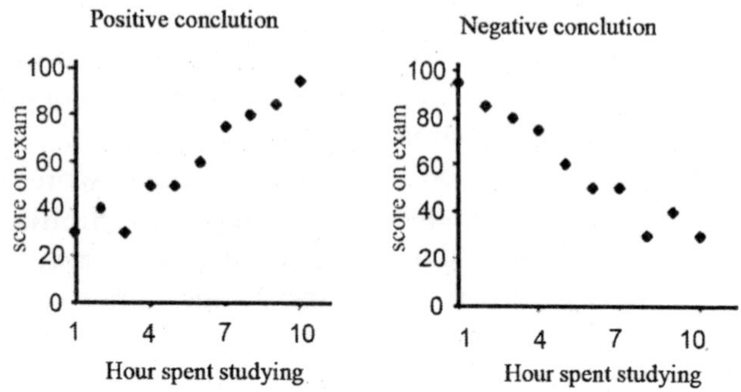

Fig. Examples of Positive and Negative Correlations.

That is, scores on one of the variables are not related in any meaningful way to scores on the second variable. The closer the correlation coefficient is to either –1.00 or +1.00, the stronger the relationship is between the two variables. A perfect negative correlation of –1.00 indicates that for every member of the sample or population, a higher score on one variable is related to a lower score on the other variable. A perfect positive correlation of + 1.00 reveals that for every member of the sample or population, a higher score on one variable is related to higher score on the other variable.

Perfect correlations are never found in actual social science research. Generally, correlation coefficients stay between –.70 and +.70. Some textbook authors suggest that correlation coefficients between –.20 and +.20 indicate a weak relation between two variables, those between.20 and.50 (either positive or negative) represent a moderate relationship, and those larger than.50 (either positive or negative) represent a strong relationship.

These general rules of thumb for judging the relevance of correlation coefficients must be taken with a grain of salt. For example, even a "small" correlation between alcohol consumption and liver disease (*e.g.*, +.15) is important, whereas a strong correlation between how much children like vanilla and chocolate ice cream (*e.g.*, +.70) may not be so important. The scattergrams presented in Figure represent very strong positive and negative correlations (*r*=.97 and *r* = ".97 for the positive and negative correlations, respectively; *r* is the symbol for the sample Pearson correlation coefficient). In Figure, a scattergram representing virtually no correlation between the number of hours spent studying and the scores on the exam is presented. Notice that there is no discernible pattern between the scores on the two variables. In other words, the data presented in Figure reveal that it would be virtually impossible to predict an individual's test score simply by knowing how many hours the person studied for the exam.

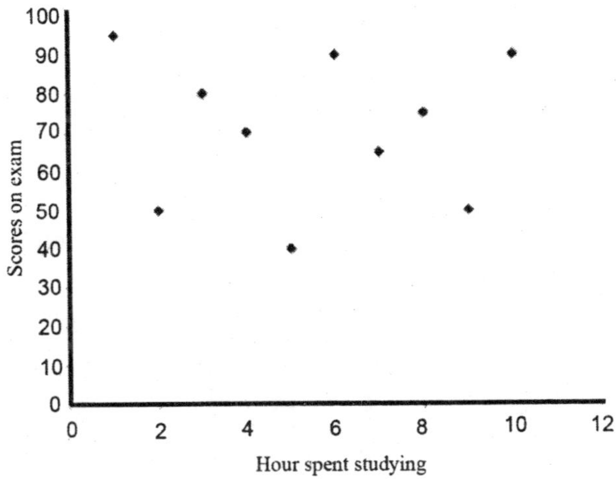

Fig. No Correlation Between Hours Spent Studying and Exam Scores

PEARSON'S CORRELATION

The correlation between two variables reflects the degree to which the variables are related. The most common measure of correlation is the Pearson Product Moment Correlation (called Pearson's correlation for short). When measured in a population the Pearson Product Moment correlation is designated by the Greek letter rho (ρ). When computed in a sample, it is designated by the letter "r" and is sometimes called "Pearson's r." Pearson's correlation reflects the degree of linear relationship between two variables. It ranges from +1 to –1. A correlation of +1 means that there is a perfect positive linear relationship between variables. The scatterplot shown on this page depicts such a relationship. It is a positive relationship because high scores on the X–axis are associated with high scores on the Y–axis. A correlation of -1 means that there is a perfect negative linear relationship between variables. The scatterplot shown below depicts a negative relationship. It is a negative relationship because high scores on the X-axis are associated with low scores on the Y-axis.

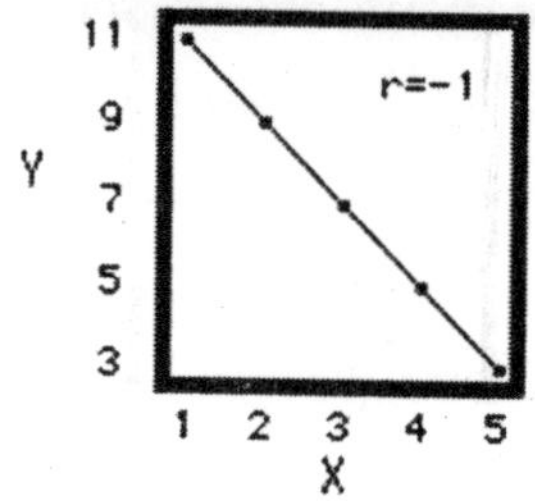

A correlation of 0 means there is no linear relationship between the two variables. The second graph shows a Pearson correlation of 0.

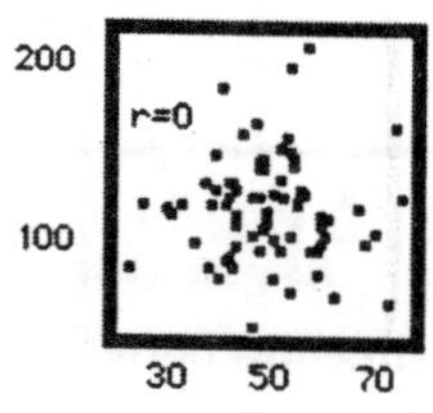

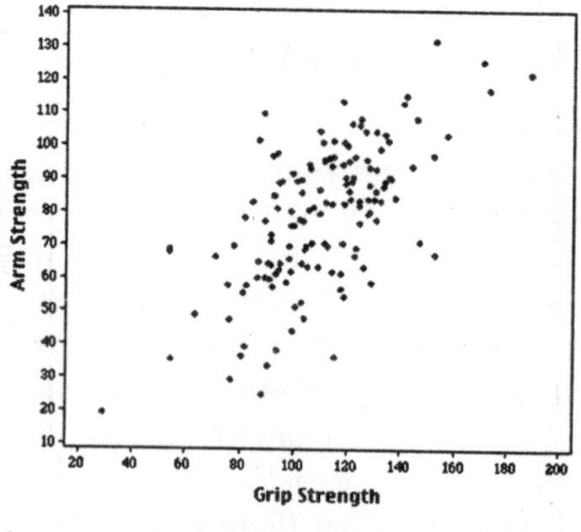

Correlations are rarely if ever 0, 1, or -1. Some real data showing a moderately high correlation are shown on the next page. The scatterplot below shows arm strength as a function of grip strength for 147 people working in physically-demanding jobs. The plot reveals a strong positive relationship. The value of Pearson's correlation is 0.63.

PEARSON CORRELATION COEFFICIENTS

The first step in understanding how Pearson correlation coefficients are calculated is to notice that we are concerned with a sample's scores on *two* variables at the same time. Returning to our previous example of study time and test scores, suppose that we randomly select a sample of five students and measure the time they spent studying for the exam and their exam scores. The data are presented in Table.

Data for Correlation Coefficient

	Hours spent Studying (X Variable)	Exam Score (Y Variable)
Student 1	5	80
Student 2	6	85
Student 3	7	70
Student 4	8	90
Student 5	9	85

For these data to be used in a correlation analysis, it is critical that the scores on the two variables *arepaired*. That is, for each student in our sample, the score on the x variable (hours spent studying) is paired with his or her own score on the *Y* variable (exam score). If we want to determine the relation between hours spent studying and exam scores, we cannot pair Student 1's hours spent studying with Student 4's test score. We must match each student's score on the x variable with his or her own score on the *Y* variable. Once we have done this, we can determine whether, on average, hours spent studying is related to exam scores.

Calculating the Correlation Coefficient

There are several different formulas that can be used to calculate Pearson product-moment correlation coefficients.

Each of these formulas produces the same result and differ only in their ease of use. In fact, none of them are particularly easy to use. We present one of them here to illustrate the principles of correlation coefficients. Should you find yourself in need of knowing the relation between two variables, it is recommend to use a calculator or statistics programme that generates correlation coefficients for you.

The formula presented in Table requires that you *standardize* your variables. Remember that when you standardize a variable, you are simply subtracting the mean from each score in your sample and dividing by the standard deviation. What this does is provide a z score for each case in the sample. Those members of the sample with scores below the mean will have negative z scores, whereas those members of the sample with scores above the mean will have positive z scores.

Definitional formula for Pearson correlation:

$$r = \frac{\Sigma\left(z_x z_y\right)}{N}$$

Wehre r = Pearson product moment correlation coefficient:

z_x = a z score for variable X

z_y = a paired z score for variable Y

N = the number of pairs of X and Y scores

Notice that this formula looks similar to some of the other formulas that we have already encountered. For example, the denominator is *N*, which is the number of pairs of scores (*i.e.*, the number of cases in the sample). Whenever we divide by *N*, we are finding an *average*.

So we know that the correlation coefficient will be an *average* of some kind. But what is it an average of? Now take a look at the numerator. Here, we see that we must find the sum (E) of something. There, we had to find the sum of the squared deviations between each individual score and the mean. But in the formula for computing the correlation coefficient, we have to find the sum of the cross products between the z scores on each of the two variables being examined for each case in the sample. When we multiply each individual's score on one variable with that individual's score on the second variable (*i.e.*, find a cross product), sum those across all of the individuals in the sample, and then divide by *N*, we have an average cross product, and this is known as covariance. If we standardize this covariance, we end up with a correlation coefficient. In the formula provided in Table, we simply standardized the variables before we computed the cross products, thereby producing a standardized covariance statistic, which is a correlation coefficient.

In this formula, notice what is happening. First, we are multiplying the paired z scores together. When we do this, notice that if an individual case in the sample has scores above the mean on each of the two variables being examined, the two z scores being multiplied will both be positive, and the resulting cross product will also be positive.

Similarly, if an individual case has scores below the mean on each of the two variables, the two z scores being multiplied will both be negative, and the cross product will again be positive.

Therefore, if we have a sample where low scores on one variable tend to be associated with low scores on the other variable, and high scores on one variable tend to be associated with high scores on the second variable, then when we add up to products from our multiplications we will end up with a *positive* number. This is how we get a positive correlation coefficient.

Now consider what happens when high scores on one variable are associated with low scores on the second variable. If an individual case in a sample has a score that is higher than the average on the first variable (*i.e.*, a positive z score) and a score that is below the mean on the second variable (*i.e.*, a negative z score), when these two z scores are multiplied together, they will produce a *negative* product.

If, for most of the cases in the sample, high scores on one variable are associated with low scores on the second variable, the sum of the products of the z scores [E(zxZy)] will be *negative*. This is how we get a negative correlation coefficient.

Correlation coefficients such as the Pearson are very

powerful statistics. They allow us to determine whether, on average, the values on one variable are *associated -with* the values on a second variable. This can be very useful information, but people, including social scientists, are often tempted to ascribe more meaning to correlation coefficients than they deserve. Namely, people often confuse the concepts of *correlation* and causation. Correlation (co-relation) simply means that variation in the scores on one variable *correspond* with variation in the scores on a second variable. Causation means that variation in the scores on one variable *cause* or *create* variation in the scores on a second variable.

When we make the leap from correlation to causation, we may be wrong. As an example, a short story is offered, which is heard in one of introductory psychology class. As the story goes, one winter shortly after World War II, there was an explosion in the number of storks nesting in some northern European country. Approximately 9 months later, there was a large jump in the number of babies that were born. Now, the link between storks and babies being what it is, many concluded that this correlation between the number of storks and the number of babies represented a causal relationship.

Fortunately, science tells us that babies do not come from storks after all, at least not human babies. However, there is something that storks and babies have in common: Both can be "summoned" by cold temperatures and warm fireplaces. It seems that storks like to nest in warm chimneys during cold winters. As it happens, cold winter nights also foster baby-making behaviour. The apparent cause-and-effect relationship between storks and babies was in fact caused by a third variable: a cold winter.

For a more serious example, we can look at the relationship between SAT scores and first-year college grade point average. The correlation between these two variables is about.40. Although these two variables are moderately correlated, it would be difficult to argue that higher SAT scores *cause* higher achievement in the first year of college. Rather, there is probably some other variable, or set of variables, responsible for this relationship. For example, we know that taking a greater number of advanced math courses in high school is associated with higher SAT scores *and* with higher grades in first-year math courses in college. The point of these examples is simple: Evidence of a relationship between two variables (*i.e.*, a correlation) does not necessarily mean that there is a causal relationship between the two variables. However, it should also be noted that a correlation between two variables is a *necessary ingredient* of any argument that the two variables are causally related. In other words, we cannot claim that one variable causes another (*e.g.*, that smoking causes cancer) if there is no correlation between smoking and cancer. If we do find a correlation between smoking and cancer, we must rule out other factors before we can conclude that it is smoking that causes cancer.

In addition to the correlation-causation issue, there are a few other important features of correlations worth noting. First, simple Pearson correlations are designed to examine *linear* relations among variables. In other words, they describe average *straight* relations among variables. For example, if you find a positive correlation between two variables, you can predict how much the scores in one variable will increase with each corresponding increase in the second variable. But not all relations between variables are linear. For example, there is a curvilinear relationship between anxiety and performance on a number of academic and nonacademic behaviours. When taking a math test, for example, a little bit of anxiety actually may help performance. However, once a student becomes too nervous, this anxiety can interfere with performance. We call this a curvilinear relationship because what began as a positive relationship between performance and anxiety at lower levels of anxiety becomes a negative relationship at higher levels of anxiety. This curvilinear relationship is presented graphically in Figure. Because correlation coefficients show the *average* relation between two variables, when the relationship between two variables is curvilinear, the correlation coefficient can be quite small, suggesting a weaker relationship than may actually exist.

	Hours spent Studying (X Variable)	Exam Score (Y Variable)
Student 1	0	95
Student 2	2	895
Student 3	4	100
Student 4	7	95
Student 5	10	100

Another common problem that arises when examining correlation coefficients is the problem of truncated range. This problem is encountered when the scores on one or both of the variables in the analysis do not have much variety in the distribution of scores, possibly due to a ceiling or floor effect.

For example, suppose that we gave a sample of students a very easy test with a possible high score of 100. Then suppose we wanted to see if there was a correlation between scores on our test and how much time students spent studying for the test. Suppose we got the following data, presented in Table.

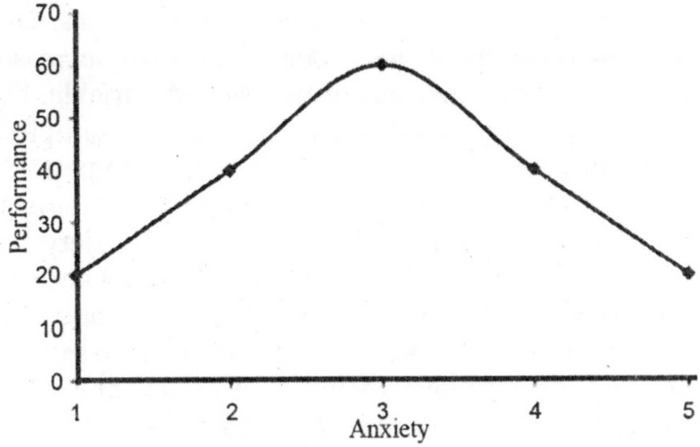

Fig. A curvilinear relationship.

In this example, all of our students did well on the test, whether they spent many hours studying for it or not. Because the test was too easy, a ceiling effect may have occurred, thereby truncating the range of scores on the exam. Although there may be a relationship between how much time students spend studying and their knowledge of the material, our test was not sensitive enough to reveal this relationship. The weak correlation that will be produced by the data in Table may not reflect the true relationship between how much students study and how much they learn.

STATISTICALLY SIGNIFICANT CORRELATIONS

When researchers calculate correlation coefficients, they often want to know whether a correlation found in sample data represents the existence of a relationship between two variables in the population from which the sample was selected. In other words, they want to test whether the correlation coefficient is statistically significant. To test whether a correlation coefficient is statistically significant, the researcher begins with the null hypothesis that there is absolutely no relationship between the two variables in the population, or that the correlation coefficient in the population equals zero.

The alternative hypothesis is that there, in fact, is a statistical relationship between the two variables in the population, and that the population correlation coefficient is *not* equal to zero. So what we are testing here is whether our correlation coefficient is statistically significantly different from 0.

These two competing hypothesis can be expressed with symbols:

$$H_{:} : \rho = 0$$

$$H_1 : \rho \neq 0$$

Where ρ is rho, the population correlation coefficient. The *t* distribution is used to test whether a correlation coefficient is statistically significant. Therefore, we must conduct a *t* test. As with all *t* tests, the *t* test that we use for correlation coefficients involves a ratio, or fraction. The numerator of the fraction is the difference between two values. The denominator is the standard error. When we want to see whether a sample correlation coefficient is significantly different from zero, the numerator of the *t* test formula will be the sample correlation coefficient, *r*, minus the hypothesized value of the population correlation coefficient (p), which in our null hypothesis is zero. The denominator will be the standard error of the sample correlation coefficient:

$$t = \frac{r - \rho}{s_r}$$

where ***r*** is the sample correlation coefficient,

ρ is the population correlation coefficient,

Fortunately, with the help of a little algebra, we do not need to calculate s_r to calculate the *t* value for correlation coefficients. First, for the sake of knowledge, let me present the formula for s_r:

$$s_r = \sqrt{\left(1 - r^2\right) \div (N - 2)}$$

where r^2 is the correlation coefficient squared and *N* is the number of cases in the sample. The formula for calculating the *t* value is,

$$t = (r)\sqrt{\frac{N-2}{1-r^2}}$$

To illustrate this formula in action, let's consider an example. Some research suggests that there is a relationship between the number of hours of sunlight people are exposed to during the day and their mood. People living at extreme northern latitudes, for example, are exposed to very little sunlight in the depths of winter days and weeks without more than a few hours of sunlight per day.

There is some evidence that such sunlight deprivation is related to feelings of depression and sadness. In fact, there is even a name for the condition: seasonal affective disorder, or SAD. To examine this relationship for ourself, 100 people is random selected from various regions of the world, measure the time from sunrise to sunset on a given day where each person lives, and get a measure of each person's mood on a scale from 1 to 10 (1="very sad, " 10="very happy").

Because the members of our sample live at various latitudes, the number of daylight hours will vary. If we conduct our study in January, those participants living in the north will have relatively short days whereas those living in the south will have long days.

Suppose that we were to compute a Pearson correlation coefficient with these data and found that the correlation between number of sunlight hours in the day and scores on the mood scale was r =.25. Is this a statistically significant correlation? To answer that question, we must find a *t* value associated with this correlation coefficient and determine that probability of obtaining a *t* value of this size by chance. In this example,

$$t = (.25)\sqrt{\frac{100-2}{1-.25^2}} \qquad t = (.25)\sqrt{\frac{98}{1-.25^2}}$$

$$t = (.25)\sqrt{\frac{98}{1-.0625}} \qquad t = (.25)\sqrt{\frac{98}{1-.9375}}$$

$$t = (.25)\sqrt{104.53} \qquad t = (.25)\ 10.22$$

$t = 1.56$, $df = 98$

To see whether this *t* value is statistically significant, we must look at the table of *t* values in Appendix B. There we can see that, because our degrees of freedom equals 98, we must look at *t* values in both the $df = 60$ row and the $df =$ 120 row. Looking at the $df = 60$ row, we can see that a *t* value of 2.56 has a probability of between.01 and.02 (for a two-tailed test). We get the same results when looking in the $df = 120$ row. Therefore, we conclude that our p value is between.01 and.02. If our alpha level is the traditional.05, we would conclude that our correlation coefficient is statistically significant. In other words, we would conclude that, on the basis of our sample statistic, in the larger population of adults the longer the daylight hours, the better their mood, on average. We could convey all of that information to the informed reader of statistics by writing "We found a significant relationship between number of daylight hours and mood (r =.25, t =2.56, p<.05)."

This example also provides a good opportunity to once again remind you of the dangers of assuming that a correlation represents a causal relationship between two variables. Although it may well be that longer days cause the average adult to feel better, these data do not prove it. An alternative causal explanation for our results is that shorter days are also associated with *colder* days whereas longer days are generally associated with *warmer* days. It may be the case that *warmth* causes better moods whereas the lack of warmth causes depression and sadness. If people had warm, short days they might be just as happy as if they had warm, long days. So remember: Just because two variables are correlated, it does not mean that one causes the other.

COEFFICIENT OF DETERMINATION

Although correlation coefficients give an idea of the strength of the relationship between two variables, they often feel a bit nebulous. If you get a correlation coefficient of.40, is that a strong relationship? Fortunately, correlation coefficients can be used to obtain a more concrete-feeling statistic: the coefficient of determination. Even better, it is easy to calculate.

When we want to know if two variables are related to each other, we are really asking a somewhat more complex question: Are the variations in the scores on one variable somehow associated with the variations in the scores on a second variable? Put another way, a correlation coefficient tells us whether we can know anything about the scores on one variable if we already know the scores on a second variable. In common statistical vernacular, what we want to be able to do with a measure of association, like a correlation coefficient, is be able to *explain* some of the variance in the scores on one variable with the scores on a second variable. The coefficient of determination tells us how much of the variance in the scores of one variable can be understood, or explained, by the scores on a second variable.

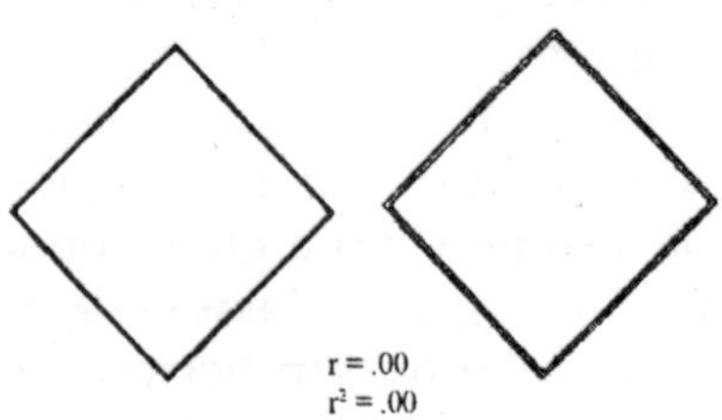

Fig. Uncorrelated variables.

One way to conceptualize explained variance is to understand that when two variables are correlated with each other, they *share* a certain percentage of their variance. Consider an example.

If we have a sample of 10 people, and we measure their height and their weight, we've got 10 scores on each of two variables. Assuming that our 10 people differ in how tall they are, there will be some total amount of variance in their scores on the height variable.

There will also be some total amount of variance in their scores on the weight variable, assuming that they do not all weigh the same amount. These total variances are depicted in Figure as two full squares, each representing 100% of the variance in their respective variables. Notice how they do not overlap.

When two variables are related, or correlated, with each other, there is a certain amount of shared variance between them. In Figure, the two squares are not touching each other, suggesting that all of the variance in each variable is independent of the other variable.

There is no overlap. But when two variables are correlated, there is some *shared* variance. The stronger the correlation, the greater the amount of shared variance, and the more variance you can explain in one variable by knowing the scores on the second variable.

The precise percentage of shared, or explained, variance can be determined by squaring the correlation coefficient. This squared correlation coefficient is known as the coefficient of determination.

Some examples of different coefficients of determination are presented in Figure. The stronger the correlation, the greater the amount of shared variance, and the higher the coefficient of determination. It is still important to remember that even though the coefficient of determination is used to tell us how much of the variance in one variable can be explained by the variance in a second variable, coefficients of determination do not necessarily indicate a *causal* relationship between the two variables.

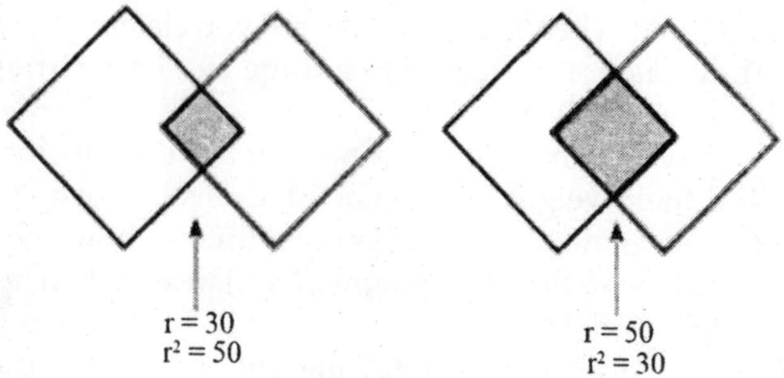

Fig. Examples of different coefficients of determination.

TYPES OF CORRELATION COEFFICIENTS

Although Pearson correlation coefficients are probably the most commonly used and reported in the social sciences, they are limited by the requirement that both variables are measured on interval or ratio scales. Fortunately, there are methods available for calculating the strength of the relationship between two variables even if one or both variables are not measured using interval or ratio scales. In this section we briefly describe three of these "other" correlation coefficients. It is important to note that all of these statistics are very similar to the Pearson correlation coefficient and each produces a correlation coefficient that is similar to the Pearson *r*. They are simply specialized versions of the Pearson correlation coefficient that can be used when one or both of the variables are not measured using interval or ratio scales.

Point Biserial

When one of our variables is a continuous variable (*i.e.*, measured on an interval or ratio scale) and the other is a two-level categorical (a.k.a. nominal) variable (also known as a dichotomous variable), we need to calculate a point-biserial correlation coefficient. This coefficient is a specialized version of the Pearson correlation coefficient discussed earlier in this chapter. For example, supposed we wanted to know whether there was a relationship between whether a person owns a car (yes or no) and their score on a written test of traffic rule knowledge, such as the tests one must pass to get a driver's license. In this example, we are examining the relation between one categorical variable with two categories (whether one owns a car) and one continuous variable (one's score on the driver's test). Therefore, the point-biserial correlation is the appropriate statistic in this instance.

Phi

Sometimes researchers want to know whether two dichotomous variables are correlated. In this case, we would calculate a phi coefficient (Ö), which is another specialized version of the Pearson *r*. For example, suppose we wanted to know whether gender (male, female) was associated with whether one smokes cigarettes or not (smoker, non smoker). In this case, with two dichotomous variables, we would calculate a phi coefficient. (Note: Those readers familiar with chi-square analysis will notice that two dichotomous variables can also be analysed using chi-square.)

Spearman Rho

Sometimes data are recorded as ranks. Because ranks are a form of ordinal data, and the other correlation coefficients discussed so far involve either continuous (interval, ratio) or dichotomous variables, we need a different type of statistic to calculate the correlation between two variables that use ranked data. The Spearman rho, a specialized form of the Pearson *r*, is appropriate. For example, many schools use students' grade point averages (a continuous scale) to rank students (an ordinal scale). In addition, students' scores on standardized achievement tests can be ranked. To see whether a students' rank in their school is related to their rank on the standardized test, a Spearman rho coefficient can be calculated.

EXAMPLE: THE CORRELATION

Between Grades and Test Scores

Student achievement can be measured in a variety of ways. One common method of evaluating students is to assign them letter grades. These grades can be converted into numbers (*e.g.*, A=4, B = 3, etc.). In high school, students grades across all of their classes (*e.g.*, mathematics, science, social studies, etc.) can be combined into an average, thereby creating a grade point average (GPA), which is measured on a continuous, interval scale ranging from a possible low of 0 to a possible high of 4.33 (if the school gives grades of A+). Because grades are assigned by teachers, they are sometimes considered to be overly subjective. That is, different teachers may assign different grades to the same work.

Similarly, there are some individual teachers that may give different grades to two students who produce the same quality of work. To ostensibly overcome such subjectivity, another form of assessment, the standardized test, was created. With this type of assessment, all students of a given grade level answer the same questions and their responses are scored by computer, thereby removing the human element and its subjectivity. Some argue that standardized tests of ability and teacher-assigned grades generally measure the same thing. That is, for the most part, bright students will score well both on the test and in their grades. Others argue that standardized tests of ability and teacher-assigned grades really measure somewhat different things. Whereas standardized tests may measure how well students answer multiple-choice questions, teachers have the benefit of knowing students, and can take things like students' effort, creativity, and motivation into account when assigning grades.

The first step in discovering which of these two viewpoints is more accurate is to see how strongly grades and test scores are related to each other. If there is a very strong correlation between the two, then both grades and test scores may in fact be measuring the same general trait. But if the two scores are only moderately correlated, perhaps they really do measure separate constructs. By constructs, we mean the actual "thing" that we are trying to measure. In the preceding example, if grades and test scores are strongly correlated, we could argue that both of these measures represent some underlying construct, such as "intelligence" or "academic ability." On the other hand, if these two variables are not strongly correlated, they each may represent different things, or constructs. Recently data is collected from a sample of 314 eleventh-grade students at a high school in California. Among the data we collected were their cumulative GPA (*i.e.*, their GPA accumulated from the time they began high school until the time the data were collected). In addition, we gave students the Naglieri Nonverbal Test of Abilities (NNTA; Naglieri, 1996), a nonverbal test of general mental reasoning and critical thinking skills.

To see if there was a statistically significant correlation between these two measures of ability, we used the SPSS statistical software programme to calculate a correlation coefficient and *a p* value. The SPSS printout from this analysis is presented in Table. The results presented in Table provide several pieces of information. First, there are three correlation coefficients presented. The correlations on the diagonal show the correlation between a single variable and itself. Therefore, the first correlation coefficient presented reveals that GPA is correlated with itself perfectly (r=1.0000). Because we always get a correlation of 1.00 when we correlate a variable with itself, these correlations presented on the diagonal are meaningless. That is why there is not a *p* value reported for them.

The numbers in the parentheses, just below the correlation coefficients, report the sample size. There were 314 eleventh-grade students in this sample. The correlation coefficient that is *off the* diagonal is the one we're interested in. Here, we can see that students' GPA was moderately correlated with their scores on the Naglieri test (*r* =.4291). This correlation is statistically significant, with *a p* value of less than.0001 (*p*<.0001). SPSS printout of correlation analysis.

	GPA	**Naglieri**
GPA	1.0000	
	(314)	
	p =.	
Naglieri	.4291	1.0000
	(314)	(314)
	p =.000	p =.

To gain a clearer understanding of the relationship between GPA and Naglieri test scores, we can calculate a coefficient of determination. We do this by squaring the correlation coefficient. When we square this correlation coefficient (.4291 ×.4291=.1841), we see that GPA explains a little bit more than 18% of the variance in the Naglieri test scores. Although this is a substantial percentage, it still leaves more than 80% of the ability test scores unexplained.

Because of this large percentage of unexplained variance, we must conclude that teacher-assigned grades reflect something substantially different from general mental reasoning abilities and critical thinking skills, as measured by the Naglieri test. Correlation coefficients, in particular Pearson correlation coefficients, provide a way to determine both the direction and the strength of the relationship between two variables measured on a continuous scale.

This index can provide evidence that two variables are related to each other, or that they are not, but does not, in and of itself, demonstrate a *causal* association between two variables. In this chapter, we also were introduced to the concepts of *explained* or *shared variance* and the coefficient of determination. Determining how much variance in one variable is shared with, or explained by, another variable is at the core of all of the statistics that are discussed in the remaining chapters of this book. In particular, correlation coefficients are the precursors to the more sophisticated statistics involved in multiple regression. We examine t tests, which allow us to look at the association between a two-category independent variable and a continuous dependent variable.

9

Regression

CONCEPT

The concept of correlation was introduced. Correlation involves a measure of the degree to which two variables are related to each other.

A closely related concept, coefficient of determination, was also introduced in that chapter. This statistic provides a measure of the strength of the association between two variables in terms of percentage of variance explained. Both of these concepts are present in regression. In this chapter, the concepts of simple linear regression and multiple regression are introduced.

Regression is a very common statistic in the social sciences. One of the reasons it is such a popular technique is because it is so versatile.

Regression, particularly multiple regression, allows researchers to examine how variables are related to each other, the strength of the relations, and relative predictive power of several independent variables on a dependent variable, and the unique contribution of one or more independent variables when controlling for one or more covariates. It is also possible to test for interactions in multiple regression.

With all of the possible applications of multiple regression, it is clear that it is impossible to describe all of the functions of regression in this brief chapter. Therefore, the focus of this chapter is to provide an introduction to the concept and uses of regression, and to refer the reader to resources to examine for additional information.

SIMPLE VERSUS MULTIPLE REGRESSION

The difference between simple and multiple regression is similar to the difference between one-way and factorial ANOVA. Like one-way ANOVA, simple regression analysis involves a single independent, or predictor variable and a single dependent, or outcome variable. This is the same number of variables used in a simple correlation analysis. The difference between a Pearson correlation coefficient and simple regression analysis is that whereas the correlation does not distinguish between independent and dependent variables, in a regression analysis there is always a designated predictor variable and a designated dependent variable. That is because the purpose of regression analysis is to make *predictions* about the value of the dependent variable given certain values of the predictor variable.

This is a simple extension of correlation analyses. If we are interested in the relationship between height and weight, for example, we could use simple regression analysis to answer this question: If we know a man's height, what would we *predict* his weight to be? Of course, the accuracy of our prediction will only be as good as our correlation will allow, with stronger correlations leading to more accurate predictions.

Therefore, simple linear regression is not really a more powerful tool than simple correlation analysis. But it does give me another way of conceptualizing the relation between two variables, a point we elaborate on shortly.

The real power of regression analysis can be found in multiple regression. Like factorial ANOVA, multiple regression involves models that have two or more predictor variables and a single dependent variable. For example, suppose that, again, we are interested in predicting how much a person weighs (*i.e.*, weight is the dependent variable). Now, suppose that in addition to height, we know how many minutes of exercise the person gets per day, and how many calories a day he consumes. Now I've got three predictor variables (height, exercise, and calories consumed) to help me make an educated guess about the person's weight.

Multiple regression analysis allows me to see, among other things:

- How much these three predictor variables, as a group, are related to weight,
- The strength of the relationship between each predictor variable and the dependent variable *while controlling for the other predictor variables in the model,*
- The *relative* strength of each predictor variable, and
- Whether there are interaction effects between the predictor variables.

As you can see, multiple regression is a particularly versatile and powerful statistical technique.

VARIABLES USED IN REGRESSION

As with correlation analysis, in regression the dependent and independent variables need to be measured on an interval or ratio scale. Dichotomous (*i.e.*, categorical variables with two categories) predictor variables can also be used. There is a special form of regression analysis, logit regression, that allows us to examine dichotomous dependent variables, but this type of regression is beyond the scope of this book. In this chapter, we limit our consideration of regression to those types that involve a continuous dependent variable and either continuous or dichotomous predictor variables.

REGRESSION IN DEPTH

Regression, particularly simple linear regression, is a statistical technique that is very closely related to correlations. In fact, when examining the relationship between two continuous (*i.e.*, measured on an interval or ratio scale) variables, either a correlation coefficient or a regression equation can be used. Indeed, the Pearson correlation coefficient is nothing more than a simple linear regression coefficient that has been standardized.

The benefits of conducting a regression analysis rather than a correlation analysis are:

- Regression analysis yields more information, particularly when conducted with one of the common statistical software packages, and

- The regression equation allows us to think about the relation between the two variables of interest in a more intuitive way.

Whereas the correlation coefficient provides us with a single number (*e.g.*, r =.40), which we can then try to interpret, the regression analysis yields a formula for calculating the predicted value of one variable when we know the actual value of the second variable. Here's how it works.

The key to understanding regression is to understand the formula for the regression equation. So we begin by presenting the most simple form of the regression equation, describe how it works, and then move on to more complicated forms of the equation. In Table, the regression equation used to find the predicted value of Y, along with definitions of the components, are presented.

Table. The Regression Equation

$$\widehat{Y} - bX + a$$

Where $\widehat{y}$ is the predicted value of the $\widehat{y}$ variable
b is the unstandardized regression coefficient, or the slope
a is intercepe (*i.e.*, the point where regression line intercepes the $\widehat{y}$ axis)

In simple linear regression, we begin with the assumption that the two variables are linearly related. In other words, if the two variables are actually related to each other, we assume that every time there is an increase of a given size in value on the X variable (called the predictor or independent variable), there is a corresponding increase (if there is a positive correlation) or decrease (if there is a negative correlation) *of a given size in* the Y variable (called the dependent, or outcome, or criterion variable). In other words, if the value of X increases from a value of 1 to a value of 2, and Y increases by 2 points, then when X increases from 2 to 3, we would predict that the value of Y would increase another 2 points.

To illustrate this point, let's consider the following set of data. Suppose we wanted to know whether there was a relationship between the amount of education people have and their monthly income. Education level is measured in years, beginning with kindergarten and extending through graduate school. Income is measured in thousands of dollars. Suppose that we randomly select a sample of 10 adults and measure then: level of education and their monthly income, getting the data provided in Table.

When we look at these data, we can see that, in general, monthly income increases as the level of education increases. This is a general, rather than an absolute, trend because in some cases a person with more years of education makes less money per month than someone with less education (*e.g.*, Case 10 and Case 9, Case 6 and Case 5). So although not every person with more education makes more money, *on average* more years of education are associated with higher monthly incomes. The correlation coefficient that describes the relation of these two variables is r=.83, which is a very strong, positive correlation.

Table: Income and Education Level Data

	Education Level (X) (In thousands)	**Monthly income(Y)**
Case 1	6 years	1
Case 2	8 years	1.5
Case 3	11 years	1
Case 3	12 years	2
Case 4	12 years	4
Case 6	13 years	5
Case 7	14 years	5
Case 8	16 years	6
Case 9	16 years	10
Case 10	21 years	8
Mean	12.9	4.1
Standard Deviation	4.25	3.32

If we were to plot these data on a simple graph, we would produce a scatterplot, such as the one provided in Figure. In this scatterplot, there are 10 data points, one for each case in the study.

Note that each data point marks the spot where education level (the *X* variable) and monthly income (the *Y* variable) meet for each case. For example, the point that has a value of 10 on the *Y* axis (income) and 16 on the *X axis* (education level) is the data point for the 10th case in our sample.

These 10 data points in our scatterplot reveal a fairly distinct trend. Notice that the points rise somewhat uniformly from the lower left corner of the graph to the upper right corner. This shape is a clear indicator of the positive relationship (*i.e.*, correlation) between education level and income. If there had been a perfect correlation between these two variables (*i.e.*, *r* " 1.0), the data points would be aligned in a perfectly straight line, rising from lower left to upper right on the graph. If the relationship between these two variables were weaker (*e.g.*, *r*=.30), the data points would be more widely scattered, making the lower-left to upper-right trend much less clear.

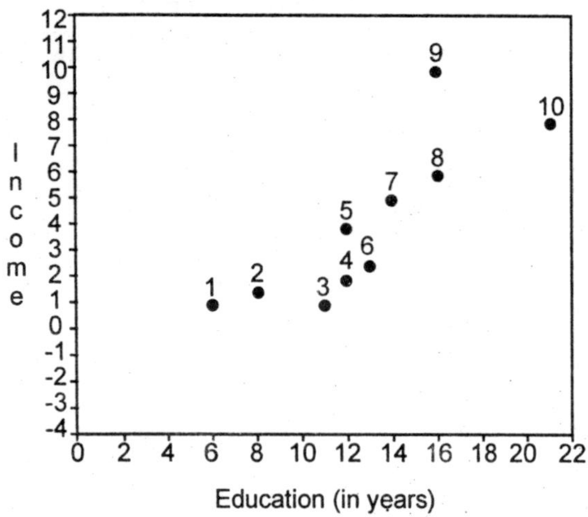

Fig. Scatterplot for education and income.

With the data provided in Table, we can calculate all of the pieces of the regression equation. The regression equation allows us to do two things. First, it lets us find predicted values for the *Y* variable for any given value of the *X* variable. In other words, we can predict a person's monthly income if we know how many years of education he or she has. Second, the regression equation allows us to produce the regression line. The regression line is the basis for linear regression and can help us understand how regression works.

There are a number of different types of regression formulas, but the most commonly used is called ordinary least squares regression, or OLS. OLS is based on an idea that we have seen before: the *sum of squares*. If you wanted to, you could draw a number of straight lines that bisect the data points presented in the scatterplot in Figure. For example, you could draw a horizontal line that extends out from the number 5 on the *Y* axis. Similarly, you could draw a straight line that extends down from the number 10 on the *Y* axis to the number 25 on the *X* axis.

No matter how you decided to draw your straight line, notice that at least some of the data points in the scatterplot will not fall exactly *on* the line. Some or all will fall above the line, some may fall directly on the line, and some or all will fall below the line. Any data point that does not fall directly on the line will have a certain amount of distance between the point and the line.

Now if you were to calculate the distance between the data point and the line you have drawn, and then square that distance, you would have a *squared deviation* for that point. If you calculated the squared deviation for each data point that did not fall on the line, and added all of these squared deviations together, you would end up with the *sum of squared deviations,* or *sum of squares.*

Now here is the key: The sum of the squared deviations, or sum of squares, will differ depending on where you draw your line. In any scatterplot, there is only *one* line the produces the *smallest* sum of squares. This line is known as the line of *least squares,* and this is the regression line.

So, the reason this type of regression is called ordinary *least squares* regression is because in this type of regression, the regression line represents the straight line that produces the *smallest* sum of squared deviations from the line. This regression line represents the *predicted* values of *Y* at any given value of X. Of course, when we predict a value of Y for a given value of X, our prediction may be off. This error in prediction is represented by the distance between the regression line and the actual data point(s) in the scatterplot. To illustrate how this works, we first need to calculate the properties of the regression line (*i.e.*, its slope and intercept). Then, we draw this regression line into the scatterplot, and you can see how well it "fits" the data (*i.e.*, how close the data points fall to the regression line).

If you take a look at the formula for the regression equation in Table, you will that there are four components:

- Y is the predicted value of the 7 variable,
- b is the unstandardized regression coefficient, and is also the slope of the regression line,
- X is the value of the X variable, and
- a is the value of the intercept (*i.e.*, where the regression line crosses the Y axis).

Because Y is the value produced by regression equation, let's save that one for last. And because X is just a given value on the X variable, there is not really anything to work out with that one. So let's take a closer look at a and b.

We cannot calculate the intercept before we know the slope of the regression line, so let's begin there. The formula for calculating the regression coefficient is,

$$b = r \times \frac{S_y}{S_x}$$

where b *is the regression coefficient,*

r *is the correlation between the X and Y variables,*

S_y *is the standard deviation of the Y variable,*

S_x *is the standard deviation of the X variable,*

Looking at the data in Table, we can see that r =.83, s_y =3.12, s_x =4.25. When we plug these numbers into the formula, we get the following:

$$b = .83 \times \frac{3.12}{4.25}$$

$$b = (.83) \times (.73)$$

$$b = .61$$

Notice that the regression coefficient is simply the correlation coefficient times the ratio of the standard deviations for the two variables involved. When we multiply the correlation coefficient by this ratio of standard deviations, we are roughly transforming the correlation coefficient into the scales of measurement used for the two variables.

Notice that there is a smaller range, or less variety, of scores on our Y variable that there is on our X variable in this example. This is reflected in the ratio of standard deviations used to calculate b. Now that we've got our b, we can calculate our intercept, a.

The formula for a is as follows:

$$a = \bar{Y} - b\bar{X}$$

where $\bar{Y}$ is the average value of Y, $\bar{X}$ is the average value of X, and b is the regression coefficient. When we plug in the values from Table, we find that,

$$a = 4.1-(.61)(12.9)$$
$$a = 4.1-7.87$$
$$a = -3.77$$

This value of a indicates that the intercept for the regression line is –3.77. In other words, the regression line crosses the Y axis at a value of –3.77. In still other words, this intercept tells us that when X=0, we wo uld predict the value of Y to be –3.77. Of course, in the real world, it is not possible to have a monthly income of negative 3.77 thousand dollars. Such unrealistic values remind us that we are dealing with *predicted* values of Y. Given our data, if a person has absolutely no formal education, we would *predict* that person to make a negative amount of money.

Now we can start to fill out our regression equation. The original formula,

$$\hat{Y} = a + bX$$

now reads,

$$\hat{Y} = -3.77 + .61X$$

It is important to remember that when we use the regression equation to find predicted values of Y for different values of X, we are not calculating the *actual* value of Y. We are only making *predictions* about the value of Y. Whenever we make predictions, we will sometimes be incorrect. Therefore, there is bound to be some error *(e)* in our predictions about the values of *Y* at given values of X. The stronger the relationship (*i.e.*, correlation) between our X and 7 variables, the less error there will be in our predictions. The error is the difference between the actual, or observed, value of *Y* and the predicted value of *Y*. Because the predicted value of *Y* is simply $a + bX$, we can express the formula for the error in two ways:

$$e = Y - \hat{Y}$$

$$e = Y - a + bX$$

So rather than a single regression equation, there are actually two. One of them, the one presented in Table, is for the *predicted* value of *Y (Y)*. The other one is for the actual, or *observed* value of *Y*. This equation takes into account the errors in our predictions, and is written as $Y=bX+a + e$.

Now that we've got our regression equation, we can put it to use. First, let's wrap words around it, so that we can make sure we understand what it tells us. Our regression coefficient tells us that "For every unit of increase in *X*, there is a corresponding predicted increase of.61 units in *Y*" Applying this to our variables, we can say that "For every additional year of education, we would *predict* an increase of.61 ($1,000), or $610, in monthly income." We know that the predicted value of *Y* will *increase* when *X* increases, and vice-versa, because the regression coefficient is *positive*. Had it been negative, we would predict a decrease in *Y* when *X* increases.

Next, let's use our regression equation to find predicted values of *Y* at given values of x. For example, what would we predict the monthly income to be for a person with 9 years of formal education? To answer this question, we plug in the value of 9 for the X variable and solve the equation:

$$\hat{Y} = -3.77 + .61(9)$$

$$\hat{Y} = -3.77 + 5.59$$

$$\hat{Y} = 1.82$$

So we would predict that a person with 9 years of education would make $1,820 per month, plus or minus our error in prediction *(e)*.

Finally, we can use our regression equation to compute our regression line. We already know, from the value of the intercept, that our regression line will cross the *Y* axis at a value of "3.77. To draw a straight line, all we need to do is calculate one additional point. To make sure we include all of the points in our scatterplot, let's just calculate a predicted value of y for a person with 25 years of education.

$$\hat{Y} = -3.77 + .61(25)$$

$$\hat{Y} = -3.77 + 15.25$$

$$\hat{Y} = 11.48$$

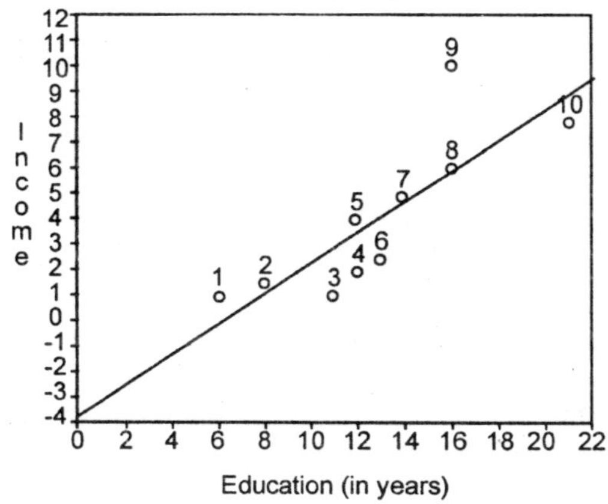

Fig. Scatterplot with Regression Line

If we were to draw a regression line through our scatterplot, using the two points we found from our intercept and the predicted *Y* value, we would get something like the line presented in Figure. With the regression line added to the scatterplot, some of the concepts mentioned earlier in this chapter may be easier to understand.

First, notice that our regression line does not accurately predict the actual y values for any of our cases except for Case 8. That data point is precisely on the regression line. For each of the other nine cases, there is some amount of error present in the prediction. In some cases, the amount of error is very little (*e.g.*, Case 7), whereas in others the amount of error is quite large (*e.g.*, Case 9). These errors in prediction are known as residuals. In some cases, our predicted value was greater than our observed value (*e.g.*, Cases 1, 2, 5, 7, and 9).

For these cases, we have overpredicted their income based on their level of education. Such overpredictions produce *negative* residuals (because the residual=observed scores-predicted score). For other cases (Cases 3, 4, 6, and 10) we underpredicted the *Y* value, creating *positive* residuals. Second, notice the distance between each case and the line. When we square each of these distances, and then add them all together, we get the sum of squares. Third, notice that the regression line marks the line where the sum of the squared distances is smallest.

To test this, try drawing some other lines and noting the way it increases the overall amount of error in prediction. Finally, notice where the regression line crosses the *Y* axis (the intercept) and the how much higher up the *Y* axis the regression line goes for each increase of one unit value in *X* (the slope). The slope and the intercept will correspond with the values that we found for *b* and a, respectively.

MULTIPLE REGRESSION

Now that we've discussed the elements of simple linear regression, let's move on to a consideration of multiple regression. Despite the impressive qualities of simple linear regression, the plain truth is that when we only have two variables, simple linear regression does not provide much more information than would a simple correlation coefficient. Because of this, you rarely see a simple linear regression with two variables reported in a published study. But multiple regression is a whole different story. Multiple regression is a very powerful statistic that can be used to provide a staggering array of useful information.

At this point, it may be worth reminding you that in a short book like this, we only scratch the surface of what multiple regression can do and how it works. The interested reader should refer to one or all of the references listed at the end of this chapter to find more information on this powerful technique.

To illustrate some of the benefits of multiple regression, let's add a second *predictor* variable to our example. So far, using the data from Table, we have examined the relationship between education level and income. In this example, education level has been used as *our predictor* or *independent* variable and income has been used as our *dependent* or *outcome* variable. We found that, on average, in our sample one's monthly salary is predicted to increase by $610 for every additional year of schooling the individual has received.

But there was some error in our predictions, indicating that there are other variables that predict how much money one makes.

One such predictor may be the length of time one has been out of school. Because people tend to make more money the longer they have been in the workforce, it stands to reason that those adults in our sample who finished school a long time ago may be making more than those who finished school more recently. Although Case 4 and Case 5 each had 12 years of schooling, Case 5 makes more money than Case 4. Perhaps this is due to Case 5 being in the workforce longer than Case 4.

When we add this second predictor variable to the model, we get the following regression equation:

$$\widehat{Y} = a + bX_1 + bX_2$$

where $\widehat{y}$ is the predicted value of the dependent variable, X_1 is the value of the first predictor variable, and X_2 is the value of the second predictor variable.

This regression equation with two predictor variables will allow me to examine a number of different questions. First, we can see whether our two predictor variables, combined, are significantly related to, or predictive of, our dependent variable, and how much of the variance our predictor variables explain in our dependent variable.

Second, we can test whether each of our predictor variables is significantly related to our dependent variable *when controlling for the other predictor variable.*

When we say "controlling for the other predictor variable" we mean that we can examine whether a predictor variable is related to the dependent variable after we partial out, or take away, the portion of the variance in our dependent variable that has already been accounted for by our other independent variable. Third, we can see which of our two predictor variables is the stronger predictor of our dependent variable. Fourth, we can test whether one predictor variable is related to our dependent variable after controlling for the other predictor variable, thus conducting a sort of ANCOVA.

Table. Income and Education Level Data

	Education Level (X_1)	Working (X_1)	Monthly income (T)
Case 1	6 years	10	1
Case 2	8 years	14	1.5
Case 3	11 years	8	1
Case 4	12 years	7	2
Case 5	12 years	20	4
Case 6	13 years	15	2.5
Case 7	14 years	17	5
Case 8	16 years	22	6
Case 9	16	30	10
Case 10	21 years	16	8
Mean	12.9	15	4.1
Standard Deviation	4.25	7.20	3.32
Concielation with income	r =.83	r =.70	

There are many other things we can do with multiple regression, but we will limit our discussion to these four. (Note: Perhaps the most important omission from this discussion is the ability to examine statistical interactions between the predictor variables in multiple regression.

Suppose that for our 10 cases in our sample we also measured the number of years that they have been in the workforce, and get the data presented in Table. The data presented in Table reveal that both years of education and years in the workforce are positively correlated with monthly income.

But how much of the variance in income can these two predictor variables explain *together?* Will years of education still predict income *when -wecontrol for the effects of years in the workforce?*

In other words, after we partial out the portion of the variance in income that is accounted for by years in the workforce, will years of education still be able to help us predict income? Which of these two independent variables will be the stronger predictor of income? And will each make a unique contribution in explaining variance in income?

To answer these questions, we use the SPSS statistical software package to analyse our data. (Note: With only 10 cases in our sample, it is not wise to run a multiple regression.

We are doing so for illustration purposes only. When conducting multiple regression analyses, you should have at least 30 cases plus 10 cases for each predictor variable in the model.)

We begin by computing Pearson correlation coefficients for all three of the variables in the model. The results are presented in Table.

Table. Correlation among Variables in Regression Model

	Years of Education	**Years in Workforce**	**Monthly Income**
Years of Education	1,000		
Years in Workforce	.310	1,000	1.00
Monthly Income	.826	.995	1.00

These data reveal that both level of education and years in the workforce are both correlated with monthly income (r =.826 and r =.695 for education and workforce with income, respectively).

In Table we can also see that there is a small-to-moderate correlation between our two predictors, years of education and years in the workforce (r=.310). Because this correlation is fairly weak, we can infer that both of these independent variables may predict education level.

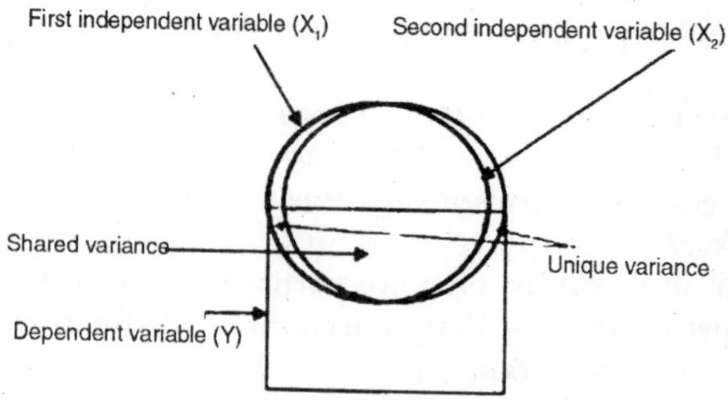

Fig. Shared Variance in Multiple Regression

Remember that in a multiple regression, we've got multiple predictor variables trying to explain variance in the dependent variable.

For a predictor variable to explain variance in a dependent variable, it must be *related* to the dependent variable. In our current example, both of our predictor variables are strongly correlated with our dependent variable, so this condition is met.

In addition, for each of our predictor variables to explain a *unique,* or *independent* portion of the variance in the dependent variable, our two predictor variables cannot be too strongly related to *each other.*

If our two predictor variables are strongly correlated with each other, then there is not going to be very much unexplained variance *in our predictor variables* left over to explain variance in the dependent variable. For example, suppose that the correlation between scores on a reading test were strongly correlated with scores on a writing test (r =.90).

Now suppose that we wanted to use reading and writing test scores to predict students' grades in the English class. Because reading and writing test scores are so highly correlated with each other, we will probably not explain any more of the variance in English class grades using both predictor variables than if we had just used one or the other. In other words, once we use reading test scores to predict English class grades, adding writing test scores to our regression model will probably not explain any more of the variance in our dependent variable, because reading and writing test scores are so closely related to each other.

This concept is represented graphically in Figure. The shaded area represents shared variance. Notice that the shaded area in the two predictor variables is so large, it is virtually impossible for any of the unshaded areas in each predictor variable to overlap with the dependent variable.

These unshaded areas represent the unique variance explaining power of each predictor. You can see that when these unique portions of the predictor variables are small, it is difficult for each predictor to explain a unique portion of the variance in the dependent variable.

Strong correlations among predictor variables is called multicollinearity and can cause problems in multiple regression analysis because it can make it difficulty to identify the unique relation between each predictor variable and the dependent variable.

CLASSIFICATION AND TABULATION

In any statistical investigation, the collection of the numerical data is the first and the most important matter to be attended.

Often a person investigating, will have to collect the data from the actual field of inquiry.

For this he may issue suitable questionnaires to get necessary information or he may take actual interviews; personal interviews are more effective than questionnaires, which may not evoke an adequate response.

Another method of collecting data may be available in publications of Government bodies or other public or private organizations.

Sometimes the data may be available in publications of Government bodies or other public or private organizations. Such data, however, is often so numerous that one's mind can hardly comprehend its significance in the form that it is shown.

Therefore it becomes, very necessary to tabulate and summarize the data to an easily manageable form. In doing so we may overlook its details. But this is not a serious loss because Statistics is not interested in an individual but in the properties of aggregates. For a layman, presentation of the raw data in the form of tables or diagrams is always more effective.

Sometimes the data may be available in publications of Government bodies or other public or private organizations. Such data, however, is often so numerous that one's mind can hardly comprehend its significance in the form that it is shown.

Therefore it becomes, very necessary to tabulate and summarize the data to an easily manageable form. In doing so we may overlook its details.

But this is not a serious loss because Statistics is not interested in an individual but in the properties of aggregates. For a layman, presentation of the raw data in the form of tables or diagrams is always more effective.

TABULATION

It is the process of condensation of the data for convenience, in statistical processing, presentation and interpretation of the information.

A good table is one which has the following requirements:

- It should present the data clearly, highlighting important details.
- It should save space but attractively designed.
- The table number and title of the table should be given.
- Row and column headings must explain the figures therein.
- Averages or percentages should be close to the data.
- Units of the measurement should be clearly stated along the titles or headings.
- Abbreviations and symbols should be avoided as far as possible.
- Sources of the data should be given at the bottom of the data.
- In case irregularities creep in table or any feature is

not sufficiently explained, references and foot notes must be given.

- The rounding of figures should be unbiased.

CLASSIFICATION

"Classified and arranged facts speak of themselves, and narrated they are as dead as mutton" This quote is given by J.R. Hicks.

The process of dividing the data into different groups which are homogeneous within but heterogeneous between themselves, is called a classification. It helps in understanding the salient features of the data and also the comparison with similar data. For a final analysis it is the best friend of a statistician.

METHODS OF CLASSIFICATION

The data is classified in the following ways:

- According to attributes or qualities this is divided into two parts:
  - Simple classification
  - Multiple classification.
- According to variable or quantity or classification according to class intervals:

Qualitative Classification: When facts are grouped according to the qualities (attributes) like religion, literacy, business etc., the classification is called as qualitative classification.

- *Simple Classification:* It is also known as classification according to Dichotomy. When data (facts) are divided into groups according to their qualities, the classification is called as 'Simple Classification'. Qualities are denoted by capital letters (A, B, C, D......) while the absence of these qualities are denoted by lower case letters (a, b, c, d.... etc.)

  For example,

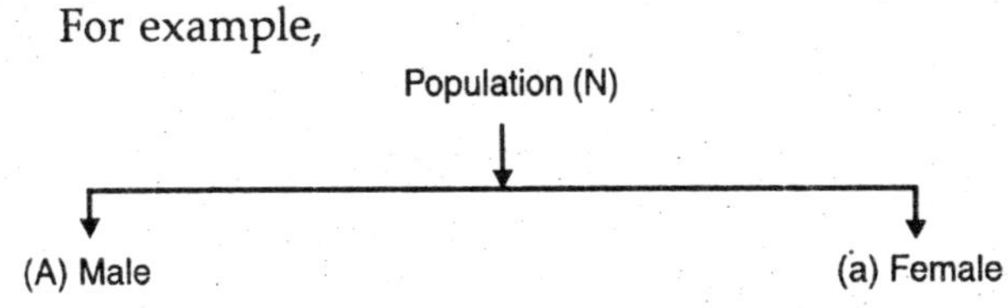

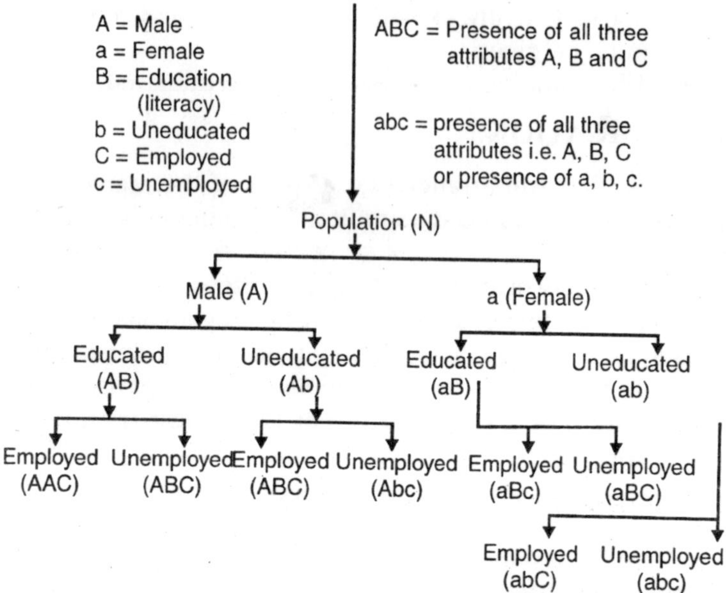

– *Manifold or multiple classification:* In this method data is classified using one or more qualities. First, the data is divided into two groups (classes) using one of the qualities. Then using the remaining qualities, the data is divided into different subgroups. For example, the population of a country is classified using three attributes: sex, literacy and business as,

Classification According to Class Intervals or Variables

The data which is expressed in numbers (quantitative data), is classified according to class-intervals. While forming class-intervals one should bear in mind that each and every item must be covered.

After finding the least value of an item and the highest value of an item, classify these items into different class-intervals.

For example if in any data the age of 100 persons ranging from 2 years to 47 years, is given, then the classification of this data can be done in this way:

Age in years	No. of persons
0-10	5
10-20	9
20-30	32
30-40	34
40-50	20

In deciding on the grouping of the data into classes, for the purpose of reducing it to a manageable form, we observe that the number of classes should not be too large.

If it were so then the object of summarization would be defeated. The number of classes should also not be too small because then we will miss a great deal of detail available and get a distorted picture. As a rule one should have between 10 and 25 classes, the actual number depending on the total frequency. Further, classes should be exhaustive; they should not be overlapping, so that no observed value falls in more than one class. Apart from exceptions, all classes should have the same length.

According to the Class-intervals in Classification the Following Terms are Used

Class-limits

A class is formed within the two values. These values are known as the class-limits of that class. The lower value is called the lower limit and is denoted by ℓ_1 while the higher value is called the upper limit of the class and is denoted by ℓ_2. In the example given above, the first class-interval has $\ell_1 = 0$ and $\ell_2 = 10$.

Magnitude of The Class-intervals

The difference between the upper and lower limits of a class is called the magnitude or length or width of a class and is denoted by 'i' or 'c'.

Thus $i \equiv (\ell_2 - \ell_1)$.

Mid-value or Class-mark

The arithmetical average of the two class limits (*i.e.* the lower limit and the upper limit) is called the mid-value or the class mark of that class-interval.

For example, the mid-value of the class-interval (0–10) is,

$$\frac{\ell_1+\ell_2}{2}=\frac{0+12}{2}=5,\ (10\text{–}20)\text{ it is }\frac{10+20}{2}=15\text{ and so on.}$$

Class frequency

The units of the data belong to any one of the groups or classes. The total number of these units is known as the frequency of that class and is denoted by *fi* or simply *f*. In the above example, the frequencies of the classes in the given order are 5, 9, 32, 34 and 40 respectively. Classification is of two types according to the class-intervals:

- Exclusive Method
- Inclusive Method

Exclusive Method

In this method the upper limit of a class becomes the lower limit of the next class. It is called ' Exclusive ' as we do not put any item that is equal to the upper limit of a class in the same class; we put it in the next class, *i.e.* the upper limits of classes are excluded from them.

For example, a person of age 20 years will not be included in the class-interval (10–20) but taken in the next class (20–30), since in the class interval (10–20) only units ranging from 10–19 are included.

The exclusive-types of class-intervals can also be expressed as:

0 and below 10 or 0–9.9

10 and below 20 or 10–19.9

20 and below 30 or 20–29.9 and so on.

Inclusive Method

In this method the upper limit of any class interval is kept in the same class-interval.

In this method the upper limit of a previous class is less by 1 from the lower limit of the next class interval. In short this method allows a class-interval to include both its lower and upper limits within it. *For example*:

Inclusive method		Inclusive method	
Class	Frequency	Class	Frequency
0–4	5	0–4.9	5
5–9	7	5–9.9	7
10–14	9	10–14.9	9
15–19	12	15–19.9	12
20–24	11	20–24.9	11
25–29	14	25–29.9	14

Class Boundaries

Weights are recorded to the nearest Kg The class-intervals 60–62 includes all measurements from 59.50000... to 62.50000... Kg; the variable being a continuous one. These numbers, indicated briefly by the exact numbers 59.5 and 62.5, are called class-boundaries or true class limits. The smaller number 59.5 is the lower class boundary and the larger one 62.5 is the upper class boundary. In any problem if the class-intervals are given as the inclusive type, then they should first be converted into the exclusive-type. For this we require a correction factor. Correction factor,

$$= \frac{1}{2}$$

(the upper limit of a class-the lower limit of the next class) which is generally 0.5. Now you subtract it from the lower limits and add it to the upper limits of the class-intervals given in the inclusive-method. The class-intervals given above can be written after correction as:

Class	Frequency
0–4.5	5
4.5–9.5	7
9.5–14.5	9
14.5–19.5	12
19.5–24.5	11
24.5–29.5	14

To obtain class-intervals when their mid-values are given, use the following formulae:

Lower limit (ℓ_1) = m–i/$_2$ and upper limit (ℓ_2) = m + i/$_2$ Where m = mid-value and i = class-length. For example, we are given some mid-values as 72, 77, 82, 87.... Now, consider the first mid-value 72 and also the differences between successive mid-values.

We have 77–72 = 5, 82–77 = 5, 87–82 = 5.... which gives the class-length i = 5. For the first class-interval, ℓ_1 = m–i/$_2$ = 72-5/$_2$ = 69.5 and ℓ_2 = 72 + 5/$_2$ = 74.5. Thus the first class-interval is 69.5–74.5 and other class-intervals then are 74.5–79.5, 79.5–84.5, 84.5–89.5

Open-end Class Intervals

In any question when the lower limit of the first class-inteval or the upper limit of the last class-interval, are not given then subtract the class length of the next immediate class-interval from the upper limit. This will give us the lower limit of the first class-interval.

Similarly add the same class length to the lower limit of the last class-interval. But always notice that the lower limit of the first class (*i.e.* the lowest class) must not be negative or less than 0.

For example:

With open ends	Completed classes	With open ends	Completed classes
Below 10	0–10	Below 10	0–10
10–20	10–20	10–25	10–25
20–30	20–30	25–40	25–40
30–40	30–40	40–70	40–70
40–50	40–50	above 70	70–100
above 50	50–60		

RELATIVE FREQUENCY DISTRIBUTION

The relative frequency of a class is the frequency of the class divided by the total number of frequencies of the class and is generally expresses as a percentage.

Example: The weight of 100 persons were given as under:

Weights (in Kgs)	60–62	63–63	66–68	69–71	72–74
No. of persons (f_i)	5	18	42	27	8

Solution:

Relative frequency table

Weight (in Kg.)	**No. of persons (f_i)**	**Relative frequency**
60–62	5	$\frac{5}{100} \times 5\%$ or 0.05
63–65	18	$\frac{18}{100} \times 18\%$ or 0.18
66–68	42	$\frac{42}{100} \times 42\%$ or 0.42
69–71	27	$\frac{27}{100} \times 27\%$ or 0.27
72–74	8	$\frac{8}{100} \times 8\%$ or 0.08
Total	$\Sigma f_i = 100$	

Note: The word frequency of a class means, the number of times the class is repeated in the data or the total number of items or observations of the data belongs to that class.

CUMULATIVE FREQUENCY

Many a times the frequencies of different classes are not given. Only their cumulative frequencies are given.

The total frequency of all values less than or equal to the upper class boundary of a given class-interval is called the cumulative frequency up to and including that class interval. In this situation both the limits of a class-interval are not written; either lower or upper limit is written. These cumulative frequencies are called less than or more than cumulative frequencies.

For example,

Class-interval	0–10	10–20	20–30	30–40	40–50
Frequency	4	9	5	12	15

(Upper limits)	**(cum. freq.)**	**(Lower limits)**	**(cum. freq.)**
Less than 10	4	More than 0	45 = 41 + 4
Less than 20	4 + 9 = 13	More than 10	41 = 32 + 9
Less than 30	13 + 5 = 18	More than 20	32 = 27 + 5
Less than 40	18 + 12 = 30	More than 30	27 = 12 + 12
Less than 50	30 + 15 = 45	More than 40	15

Preparation of Frequency Distribution

We shall now study how to classify the raw data in a tabular form. Consider the data collected by one of the surveyors, interviewing about 50 people. This is as follows: Size of the shoes: 2, 5, 6, 8, 2, 5, 6, 7, 6, 8, 7, 4, 3.. This is called the raw data. Here some values repeat themselves. For instance the size 5 is repeated 10 times in 50 people. We say that the value of 5 of the variate has the frequency of 10. Frequency means the number of times a value of the variate or an attribute, as the case may be, is repeated in the data. A table which shows each value of the characteristic with its corresponding frequency, is known as a Frequency Distribution.

Discrete variate

Consider the raw data which gives the size of shoes of 30 persons:

2, 5, 6, 4, 5, 7, 4, 4, 6, 2

3, 5, 5, 4, 5, 6, 5, 4, 3, 2

4, 4, 5, 4, 5, 5, 3, 2, 4, 4

The least value is 2 and the highest is 7. All sizes are integers between 2 and 7 (both inclusive). We can prepare a frequency distribution table as follows:

Sizes of shoes	**Tally Marks**	**Frequency**
2	||||	4
3	|||	3
4	𝍸 𝍸	10

5	卌 卌	9
6	III	3
7	I	1
Total		**30**

In this example the size difference from 2 to 7 is very small. If the range of a variate is very large, it is inconvenient to prepare a frequency distribution for each value of the variate. In such a case we divide the variate into convenient groups and prepare a table showing the groups and their corresponding frequencies. Such a table is called a grouped frequency distribution.

Consider the marks (out of 100) of 50 students as below:

40, 39, 43, 62, 30, 47, 33, 31, 17, 28
36, 29, 40, 32, 39, 24, 57, 42, 15, 30
50, 52, 47, 65, 31, 07, 37, 47, 17, 20
25, 53, 65, 85, 89, 56, 55, 41, 43, 10
44, 40, 69, 22, 40, 65, 39, 36, 71, 12

The range of the variate (marks) is very large. Also we are eager to know the performance of the students. The passing limit is 35 and above. Marks between 35 and 44 form the third class (or grade). Marks ranging between 45-59 are considered as second class and 60–100 form the first class.

Marks	Tally Marks	Frequency
0–34	卌 卌 卌 I	16
35–44	卌 卌 卌 III	18
45–59	卌 IIII	9
60–100	卌 II	7
Total		**50**

BIVARIATE FREQUENCY DISTRIBUTION

In table, we have classified students according to their marks in one subject. Suppose we have marks of a number of students in two subjects, say statistics and physics and we want to consider the variation of marks in both the subjects simultaneously.

Then we have to prepare a table (known as a bivariate table) for the following data:

No. of students	1	2	3	4	5	6	7	8	9	10	11	12
Marks in statistics	60	56	42	33	40	13	27	12	19	50	23	16
Marks in physics	50	45	53	54	49	23	50	24	31	41	47	20

No. of students	13	14	15
Marks in statistics	32	42	30
Marks in physics	38	21	45

Marks in Statistics	Marks in Physics					
	10–20	20–30	30–40	40–50	50–60	
10–20		| | = 2	| = 1			3
20–30		| = 1		| = 1	| = 1	3
30–40			| = 1	| = 1	| = 1	3
40–50		| = 1		| = 1	| = 1	3
52–60				| = 1	| | = 2	3
Total		**4**	**2**	**4**	**5**	**15**

Procedure

We have to consider a pair of data of observations. We mark one stroke (the tally mark) for one pair. Now consider the case of the 1st student. Marks in statistics = 60 Marks in physics = 50. See the row along 50–60 → 5th row and the column along 50–60–5th column. Therefore the stroke is marked in the cell (box) where the 5th row and the 5th column intersect each other. In this way we mark the strokes for 15 pairs of marks for 15 students and we get the table.

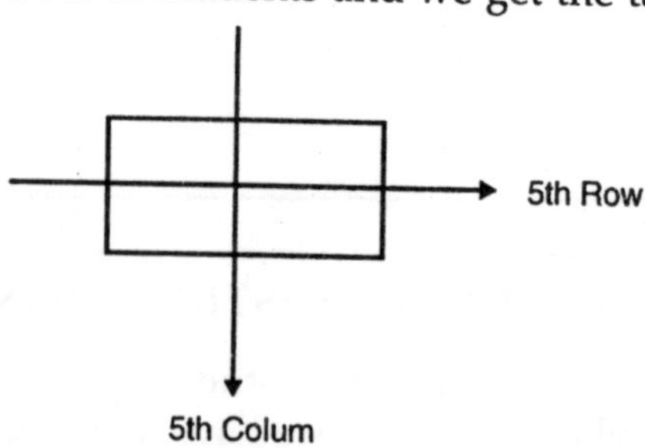

Example: Make a table showing the frequencies, cumulative frequencies (both less than and more than type) for the words having different numbers of letters in the following passage.

"In Asia the complications are likely to be greater still. There will be four major powers in the Asian balance: Russia, China, Japan and the United States. To some extent, this increases the chance that the forces will emerge sufficient to contain China and it may lead to a real reapproachment between Russia and the West.

But who knows, with so many factors, what the pattern will be like, ten years from now ?"

Solution: We find that the smallest word ' a ' has one letter; and the longest word "re-approachment" has 14 letters. We prepare a tally sheet by writing the number of letters in the first column from 1 to 14.

The table is as follows:

Number of letters	Tally-marks	Total
1	I	1
2	IIII IIII I	11
3	IIII IIII IIII I	16
4	IIII IIII IIII I	16
5	IIII IIII	9
6	IIII IIII	10
7	IIII I	6
8	–	0
9	I	1
10	I	1
11	–	0
12	–	0
13	I	1
14	I	1
Total		**73**

Number of letters (xi)	Frequency	Cumulative frequency (< or = type)	Cumulative frequency (> or = type)
1	1	1	73
2	11	12	72
3	16	28	71
4	16	44	71
5	9	53	71
6	10	63	70
7	6	69	69
8	0	69	69
9	1	70	63
10	1	71	53
11	0	71	44
12	0	71	28
13	1	72	12
14	1	73	1
Total	73		

Example: The following gives the weights (to the nearest pounds) of 40 students of the 7th grade.

Construct a frequency distribution table giving frequencies and cumulative frequencies of two types. Tabulate the data into equal classes, the 1st class being 118–122. Also indicate the class boundaries and the class-marks.

144, 125, 149, 157, 132, 150, 164, 138,

146, 158, 140, 147, 168, 126, 138, 176,

136, 148, 142, 144, 163, 119, 154, 165,

146, 173, 161, 145, 135, 142, 142, 147,

135, 153, 140, 135, 150, 156, 128, 145

Solution: The successive classes are 118–122, 123–127....., 173–177. The lower and upper class boundaries for the 1st class will be 117.5 and 122.5 and the class-mark will be

$$\frac{177.5+122.5}{2}=\frac{118+122}{2}=120$$

The class-boundaries and the class-marks will go on increasing by 5.

10

Central Tendency

STUDY

A single value which can be considered as typical or representative of a set of observations and around which the observations can be considered as Centred is called an 'Average' (or average value) or a Centre of location. Since such typical values tends to lie centrally within a set of observations when arranged according to magnitudes, averages are called measures of central tendency.

In fact the distribution have a typical value (average) about which, the observations are more or less symmetrically distributed. This is of great importance, both theoretically and practically. Dr. A.L. Bowley correctly stated, "Statistics may rightly be called the science of averages."

The word average is commonly used in day-to-day conversations. For example, we may say that Abert is an average boy of my class; we may talk of an average American, average income, etc. When it is said, "Abert is an average student," it means is that he is neither very good nor very bad, but a mediocre student. However, in statistics the term average has a different meaning.

The fundamental measures of tendencies are:

- Arithmetic mean
- Median
- Mode
- Geometric mean
- Harmonic mean
- Weighted averages

However the most common measures of central tendencies or Locations are Arithmetic mean, median and mode. We therefore, consider the Arithmetic mean.

ARITHMETIC MEAN

This is the most commonly used average which you have also studied and used in lower grades. Here are two definitions given by two great masters of statistics. Horace Sacrist: Arithmetic mean is the amount secured by dividing the sum of values of the items in a series by their number. W.I. King: The arithmetic average may be defined as the sum of aggregate of a series of items divided by their number. Thus, the students should add all observations (values of all items) together and divide this sum by the number of observations (or items).

Ungrouped Data

Suppose, we have 'n' observations (or measures) x_1, x_2, x_3......., x_n then the Arithmetic mean is obviously,

$$\frac{x_1 + x_2 + x_3 + ... + x_n}{n}$$

We shall use the symbol x (pronounced as x bar) to denote the Arithmetic mean. Since we have to write the sum of observations very frequently, we use the usual symbol 'Σ' (pronounced as sigma) to denote the sum. The symbol x_i will be used to denote, in general the 'i' th observation. Then the sum, $x_1 + x_2 + x_3 + ... + x_n$ will be represented by or simply ,

$$\sum_{i=1}^{n} x_i \text{ or } \sum x_i$$

Therefore the Arithmetic mean of the set $x_1 + x_2 + x_3 + ... + x_n$ is given by,

$$\overline{x} = \frac{\sum x_i}{n} = \frac{x_1 + x_2 + x_3 + ... + x_n}{n}$$

This method is known as the "Direct Method".

Example: A variable takes the values as given below. Calculate the arithmetic mean of 110, 117, 129, 195, 95, 100, 100, 175, 250 and 750.

Solution: Arithmetic mean = $\frac{\sum x_i}{n}$

$\sum x_i = 110 + 117 + 129 + 195 + 95 + 100 + 100 + 175 + 250 + 750$
= 2021 and $n = 10$

Indirect Method (Assumed Mean Method)

$\bar{u} = \frac{\sum u_i}{n}$ where $u = x_i - A$

A = Assumed Mean = $A + \bar{u}$

Calculations: Let $A = 175$ then

Σu_i = –65, –58, –46, +20, –80, –75, –75, +0, +75, +575

= 670–399 = 271/10 = 27.1 $\bar{x} = A + \bar{u}$ = 175 + 27.1 = 202.1

Example: M.N. Elhance's earnings for the past week were:

Monday	\$ 450
Tuesday	\$ 375
Wednesday	\$ 500
Thursday	\$ 350
Friday	\$ 270

Find his average earning per day.

Solution:

$\sum x_i$ = 450 + 375 + 500 + 350 + 270 = \$ 1945 $n = 5$

Arithmetic mean = $\frac{\sum x_i}{n} = \frac{1945}{5}$ = \$389

Therefore, Elhance's average earning per day is \$389.

PROPERTIES OF ARITHMETIC MEAN

- The sum of the deviations, of all the values of x, from their arithmetic mean, is zero.

  Justification:

  $\sum f_i(x_i - \bar{x}) = \sum f_i x_i - \bar{x}\sum f_i = 0$

  Since $\bar{x}$ is a constant,

  $\bar{x} = \frac{\sum f_i x_i}{\sum f_i} = \sum f_i x_i - = \bar{x}\sum f_i$

- The product of the arithmetic mean and the number of items gives the total of all items.

Justification:

$$\bar{x} = \frac{\sum f_i x_i}{\sum f_i} \Rightarrow \sum f_i x_i - = \bar{x} \sum f_i$$

$$\text{or } \bar{x} = \frac{\sum x}{N} \Rightarrow \bar{x}.N = \sum x_i$$

- If $\bar{x}_1$ and $\bar{x}_2$ are the arithmetic mean of two samples of sizes n_1 and n_2 respectively then, the arithmetic mean $\bar{x}$ of the distribution combining the two can be calculated as

$$\bar{x} = \frac{n_1\bar{x}_1 + n_2\bar{x}_2}{n_1 + n_2}$$

This formula can be extended for still more groups or samples.

$$\bar{x}_1 = \frac{\sum x_{1i}}{n_1} \Rightarrow \sum x_{1i} = n_1\bar{x}_1$$

Justification: $\bar{x}_1 = \dfrac{\sum x_{1i}}{n_1} \Rightarrow \sum x_{1i} = n_1\bar{x}_1$ = total of the observations of the first sample

Similarly $\sum x_{2i} = n_2\bar{x}_2$ = total of the observations of the first sample The combined mean of the two samples

$$= \frac{\text{combined total}}{n_1 + n_2}$$

$$\bar{x} = \frac{n_1\bar{x}_1 + n_2\bar{x}_2}{n_1 + n_2}$$

Example: The mean of a certain number of observations is 40. If two or more items with values 50 and 64 are added to this data, the mean rises to 42. Find the number of items in the original data.

Solution: Let 'n' be the number of observations whose mean $\bar{x} = 40$.

$$\bar{x} = \frac{\sum x_i}{n} \quad \sum x_i = n\bar{x} = n(40) = 40n \text{ total of n values.}$$

Two more items of values 50 and 64 are added therefore, total of (n + 2) values:

$$= \sum x_i + 50 + 64$$
$$= 40n + 50 + 64$$
$$= 40n + 114$$

Now new mean is 42.

$$\text{New } \bar{x} = \frac{\text{New total of (n+2) value}}{n+2}$$

$$42 = \frac{40n+14}{n+2}$$

$$42n + 84 = 40n + 114$$
$$2n = 30$$
$$n = 15$$

Therefore, the number of items in the original data = 15.

Example: The sum of deviations of a certain numbers of observations measured from 4 is 72 and the sum of deviations of observations measured from 7 is –3. Find the number of observations and their mean.

Solution:

Let 'n' be the required number of observations,

$$\sum (x_i - 4) = 72,$$

therefore,

$$\sum x_i - 4_n = 72$$

Note,

$$\sum 4 = 4n \text{ and } \sum (x_i - 7) = -3,$$

therefore,

$$\sum x_i - 7n = -3$$

Subtracting the two equations we get,

$$\sum x_i - 4n = 72$$
$$\sum x_i - 4n = -2$$
$$3n = 72$$
$$n = 25$$

Putting $n = 25$ in $\sum x_i - 4n = 72$, we get,

$$\sum x_i - 4(25) = 72$$
$$\sum x_i = 72 + 100$$
$$\sum x_i = 172$$

Now Mean is given by,

$$\bar{x} = \frac{\sum x_i}{n} = \frac{172}{25} = 6.88$$

Example: The mean weight of 98 students is found to be 50 lbs. It is later discovered that the frequency of the class interval (30- 40) was wrongly taken as 8 instead of 10. Calculate the correct mean.

Solution: Incorrect mean,

$$(\bar{x}) = 50 \text{ Kg and } \sum f_i = 98$$

$$\text{Incorrect } \bar{x} = \frac{\text{Incorrect} \sum f_i x_i}{\sum f_i}$$

$$50 = \frac{\text{Incorrect} \sum f_i x_i}{98}$$

Therefore, Incorrect $\sum f_i x_i = 98 \times 50 = 4900$

Now correct,

$$\sum f_i x_i = \text{Incorrect} \sum f_i x_i - (8 \times 35) + (10 \times 35)$$

Note that the class-mark of class interval (30-40) is 35 and for the calculation of the mean we consider class marks.

The correct $\sum f_i x_i$ = 4900–280 + 350 = 4970

Also the correct $\sum f_i$ = 98 + 2 = 100

Therefore, the correct mean

$$\bar{x} = \frac{\text{correct} \sum f_i x_i}{\text{correct} \sum f_i} = \frac{4970}{100}$$

$\bar{x}$ = 49.701b

Example: The sum of the deviations of 'n' observation values of a variate from a constant 'a', is S. Show that the arithmetic mean is,

$$a + \frac{S}{n}.$$

Solution: Dividing by n to get the mean

$$\text{i.e. } \frac{\sum x_i}{n} - \frac{an}{n} = \frac{S}{n} \quad \frac{\sum x_i}{n} = a + \frac{S}{n}$$

But,

$$\frac{\sum x_i}{n} = \bar{x}$$

Therefore, $\bar{x} = a + \dfrac{S}{n}$

Merits

- It is rigidly defined. Its value is always definite.
- It is easy to calculate and easy to understand. Hence it is very popular.
- It is based on all the observations; so that it becomes a good representative.
- It can be easily used for comparison.
- It is capable of further algebraic treatment such as finding the sum of the values of the observations, if the mean and the total number of the observations are given; finding the combined arithmetic mean when different groups are given etc.
- It is not affected much by sampling fluctuations.

Demerits

- It is affected by outliers or extreme values. For example, the average (A.) mean of 10, 15, 25 and 500 is,

$$\bar{x} = \frac{10+15+25+500}{4} = 137.5$$

Now observe first three values whose A.mean is

$$\frac{10+15+25}{3} = 16.67 \text{ (approx.)}$$

Due to the outlier 500 the A. mean of the four numbers is raised to 137.5. In such a case A. mean is not a good representative of the given data.

- It is a value which may not be present in the given data.
- Many a times it gives absurd results like 4.4 children per family.
- It is not possible to take out the averages of ratios and percentages.
- We cannot calculate it when open-end class intervals are present in the data.

MEDIAN

It is the value of the size of the central item of the arranged data (data arranged in the ascending or the descending order). Thus, it is the value of the middle item and divides the series in to equal parts.

In Connor's words-"The median is that value of the variable which divides the group into two equal parts, one part comprising all values greater and the other all values lesser than the median." For example, the daily wages of 7 workers are 5, 7, 9, 11, 12, 14 and 15 dollars. This series contains 7 terms. The fourth term *i.e.* \$11 is the median.

Median In Individual Series (ungrouped Data)

- Set the individual series either in the ascending (increasing) or in the descending (decreasing) order, of the size of its items or observations.
- If the total number of observations be 'n' then

A. If 'n' is odd,

$$\text{The median} = \text{size of } \left(\frac{n+1}{2}\right)^{\text{th}} \text{ observation}$$

B. If 'n' is even, the median

$$= \frac{1}{2}\left[\begin{array}{c} \text{size of}\left(\frac{n}{2}\right)^{\text{th}} \text{ observations} \\ +\text{size of}\left(\frac{n+2}{2}\right)^{\text{th}} \text{ observations} \end{array}\right]$$

Example: The following figures represent the number of books issued at the counter of a Statistics library on 11 different days. 96, 180, 98, 75, 270, 80, 102, 100, 94, 75 and 200. Calculate the median.

Solution: Arrange the data in the ascending order as 75, 75, 80, 94, 96, 98, 100, 102,180, 200, 270.

Now the total number of items 'n'= 11 (odd)

Therefore, the median = size of $\left(\frac{n+1}{2}\right)^{th}$ item

$= \text{size of } \left(\frac{11+1}{2}\right)^{th} \text{ item}$

= size of 5th item

= 98 books per day

Example: The population (in thousands) of 36 metropolitan cities are as follows: 2468, 591, 437, 20, 213, 143, 1490, 407, 284, 176, 263, 19, 181, 777, 387, 302, 213, 204, 153, 733, 391, 176 178, 122, 532, 360, 65, 260, 193, 92, 672, 258, 239, 160, 147, 151. Calculate the median.

Solution: Arranging the terms in the ascending order as:

20, 65, 92, 131, 142, 143, 147, 151, 153, 160, 169, 176, 178, 181, 193, 204, (213, 39), 258, 263, 260, 384, 302, 360, 387, 391, 407, 437, 522, 591, 672, 733, 777, 1490, 2488.

Since total number of items n = 36 (Even). the median

$$= \frac{1}{2}\left[\text{size of}\left(\frac{n}{2}\right)^{th} \text{item} + \text{size of}\left(\frac{n+2}{2}\right)^{th} \text{item}\right]$$

$$= \frac{1}{2}\left[\text{size of } 18^{th} \text{ item} + \text{size of} 19\text{th item}\right]$$

$$= \frac{1}{2}[213 + 239]$$

$$= \frac{1}{2}[252] = 226 \text{ thousands}$$

Median In Discrete Series

Steps:

- Arrange the data in ascending or descending order of magnitude.
- Find the cumulative frequencies.
- Apply the formula:

A. If 'n' = $\sum f_i$ (odd) then,

$$\text{Median} = \text{size of } \left(\frac{n+1}{2}\right)^{\text{th}} \text{ item}$$

B. If 'n' = $\sum f_i$ (even) then,

$$\text{Median} = \frac{1}{2}\left[\begin{array}{l}\text{size of}(n/2)^{\text{th}} \text{ item} \\ +\text{size of}\left(\frac{n+2}{2}\right)^{\text{th}} \text{ item}\end{array}\right]$$

Example: Locate the median in the following distribution.

Size:	8	10	12	14	16	18	20
Frequency:	7	7	12	28	10	9	6

Solution:

Size (x_i)	**Frequency** f_i	**Cumulative frequency** (c. f)
8	3	3
10	7	3 + 7 = 10
12	12	12 + 10 = 22
14	28	28 + 22 = 50
16	10	50 + 10 = 60
18	9	60 + 9 = 69
20	6	69 + 6 = 75
	$n = \Sigma f_i = 75$ (odd)	

Therefore, the median = $\text{size of}\left(\frac{n+1}{2}\right)^{\text{th}}$ item

$= \text{size of}\left(\frac{75+1}{2}\right)^{\text{th}}$ item

= size of 38th item

In the order of the cumulative frequency, the 38th term is present in the 50th cumulative frequency, whose size is 14. Therefore, the median = 14

Median In Continuous Series (grouped Data)

Steps:

- Determine the particular class in which the value of the median lies.

Use as the rank of the median and not,

$$\left(\frac{n+1}{2}\right)$$

- After ascertaining the class in which median lies, the following formula is used for determining the exact value of the median.

$$\text{Median} = \ell_1 + \left[\frac{N/2 - c.f}{f}\right](\ell_2 - \ell_1)$$

where,

ℓ_1 = Lower limit of the median class, the class in which the middle item of the distribution lies.

ℓ_2 = Upper limit of the median class

$c.f$ = Cumulative frequency of the class preceding the median class

f = Sample frequency of the median class

It should be noted that while interpolating the median value of frequency distribution it is assumed that the variable is continuous and that there is an orderly and even distribution of items within each class.

Example: Calculate the median for the following and verify it graphically.

Age (years):	20-25	25-30	30-35	35-40	40-45
No. of person:	70	80	180	150	20

Solution:

Age (years) C.I.	**No. of persons f_i**	**Cumulative frequency c.f.**
20-25	70	70
25-30	80	70 + 80 = 150
30-35	180	150 + 180 = 330
35-40	150	330 + 150 = 480
40-45	20	480 + 20 = 500
	$n = \Sigma fi = 500$	

Now median = size of $(n/2)^{th}$ item

$$= \text{size of } \left(\frac{500}{2}\right)^{th}$$

= size of 250th item which

lies in (30-35) class interval,

$$\ell_1 + \left[\frac{N/2 - c.f}{f}\right](\ell_2 - \ell_1)$$

Here $\ell_1 = 30$, $\ell_2 = 35$, $= n/2\ 250$, $c.f. = 150$ and $f = 180$

Therefore, Median,

$$= 30 + \left[\frac{250 - 150}{180}\right](35 - 30)$$

$$= 30 + \frac{10}{180} \times 5$$

= 32.78 years

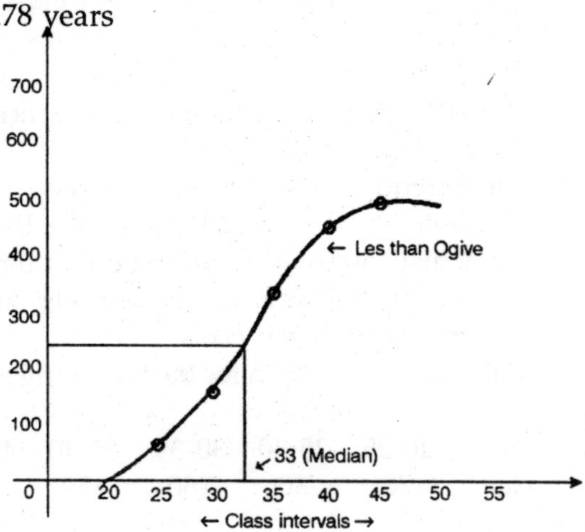

Sometimes the series is given in the descending order of magnitude. In this situation convert the series in the ascending order of magnitude and then using the regular formula, the median can be calculated or the series can be put in the descending order of the magnitude and an alternative formula be used to calculate the median.

Example:

Marks:	40-50	30-40	20-30	10-20	0-10
No. of students:	10	12	40	30	8

Solution:

Marks	**No. of students**	**c.f**
0-10	8	8
10-20	30	8 + 30 = 38

20-30	40	38 + 40 = 78
30-40	12	78 + 12 = 90
40-50	10	90 + 10 = 100
	$n = \Sigma f = 100$	

Median = size of $(n/2)^{th}$ item

$$= \text{size of } \left(\frac{500}{2}\right)^{th} \text{ item}$$

= size of 50th item which comes into class interval (20-30).

By interpolation,

$$= 20 + \left[\frac{50-38}{40}\right](30-20)$$

$$= 20 + \frac{12}{40} \times 10$$

$$= 20 + 3$$

$$\text{Median} = \ell_1 + \left[\frac{n/2 - c.f}{f}\right](\ell_2 - \ell_1)$$

Median = 23 marks

Marks	No. of students	Cumulative Frequency
40-50	10	10
30-40	12	22
20-30	40	62
10-20	30	92
0-10	8	100
	$n = \Sigma f = 100$	

Arranging the series in the descending order (as it is given) Median = size of $(n/2)^{th}$ item = size of 50th item which lies in (20-30) class-interval.

Alternative Formula

$$\text{Median} = \ell_1 - \left[\frac{n/2 - c.f}{f}\right](\ell_2 - \ell_1)$$

$$= 30 - \left[\frac{50 - 22}{40}\right](30 - 20)$$

$$= 30 - \frac{28}{40} \times 10 = 23 \text{ marks}$$

Note that, while calculating the median of a series, it must be put in the 'exclusive class-interval' form. If the original series is in inclusive type, first convert it into the exclusive type and then find its median.

Example: The following distribution represents the number of minutes spent by a group of teenagers in watching movies. What is the median?

Minutes/Weeks:0-99 100-199 200-299 300-399 400-499 500-599 600 and more

No. of teenagers: 27 32 65 78 58 32 8

Solution:

Minutes/ weeks	Real class intervals	Frequency	Cumulative frequency
0-99	0.5-99.5	27	27
100-199	99.5 199.5	32	$27 + 32 = 59$
200-299	199.5-299.5	65	$59 + 65 = 124$
300-399	299.5-399.5	78	$124 + 78 = 202$
400-499	399.5-499.5	58	$202 + 58 = 260$
500-599	499.5-599.5	32	$260 + 32 = 292$
600 and more	599.5 and more	82	$92 + 8 = 300$
		$n = \Sigma f_i = 300$	

Median = size of $(n/2)^{th}$ item

$$= \text{size of } \left(\frac{300}{2}\right)^{th} \text{ item}$$

= size of 150th item which

lies in (299.5–399.5) class interval

By using interpolation

$$\text{Median} = \ell_1 - \left[\frac{n/2 - c.f}{f}\right](\ell_2 - \ell_1)$$

$$= 299.5 + \left[\frac{150 - 124}{78}\right](399.5 - 299.5)$$

$$= 299.5 + \left[\frac{26}{78}\right] \times 100$$

299.5 + 33.33 (approximately)

Median = 392.88 minutes/week

Merits of Median

- It is rigidly defined.
- It is easy to calculate and understand.
- It is not affected by extreme values like the arithmetic mean. For example, 5 persons have their incomes \$2000, \$2500, \$2600, \$3000, \$5000. The median would be \$2600 while the arithmetic mean would be \$3020.
- It can be found by mere inspection.
- It is fully representative and can be computed easily.
- It can be used for qualitative studies.
- Even if the extreme values are unknown, median can be calculated if one knows the number of items.
- It can be obtained graphically.

Demerits of Median

- It may not be representative if the distribution is irregular and abnormal.
- It is not capable of further algebraic treatment.
- It is not based on all observations.
- It is affected by sample fluctuations.
- The arrangement of the data in the order of magnitude is absolutely necessary.

MODE

It is the size of that item which possesses the maximum frequency. According to Professor Kenney and Keeping, the value of the variable which occurs most frequently in a distribution is called the mode. It is the most common value. It is the point of maximum density.

Ungrouped Data

Individual series: The mode of this series can be obtained by mere inspection. The number which occurs most often is the mode.

Example: Locate mode in the data 7, 12, 8, 5, 9, 6, 10, 9, 4, 9, 9

Solution: On inspection, it is observed that the number 9 has maximum frequency. Therefore 9 is the mode.

Note that if in any series, two or more numbers have the maximum frequency, then the mode will be difficult to calculate. Such series are called as Bi-modal, Tri-modal or Multi-modal series.

Grouped Data

Steps:

- Determine the modal class which as the maximum frequency.
- By interpolation the value of the mode can be calculated as -

$$\text{Mode} = \ell_1 - \left[\frac{f_1 - f_0}{2f_1 - f_0 - f_2}\right](\ell_2 - \ell_1)$$

where,

ℓ_1 = Lower limit of the modal class
ℓ_2 = Upper limit of the modal class
f_1 = Frequency of the modal class
f_0 = Frequency of the class preceding to the modal class
f_2 = Frequency of the class succeding the modal class

Example: Calculate the modal wages.

Daily wages in \$:	20 -25	25-30	30-35	35-40	40-45	45-50
No. of workers:	1	3	8	12	7	5

Verify it graphically.

Solution: Here the maximum frequency is 12, corresponding to the class interval (35-40) which is the modal class.

Therefore $\ell_1 = 35,\ \ell_2 = 40,\ \ell_0 = 8,\ \ell_1 = 12,\ \ell_2 = 7,$
By interpolation

$$\text{Mode} = \ell_1 - \left[\frac{f_1 - f_0}{2f_1 - f_0 - f_2}\right](\ell_2 - \ell_1)$$

$$= 40 + \left[\frac{12-8}{24-8-7}\right](40-35)$$

$$= 40 + \left[\frac{4}{9}\right] \times 5$$

$$= 40 + 2.22 = 37.22$$

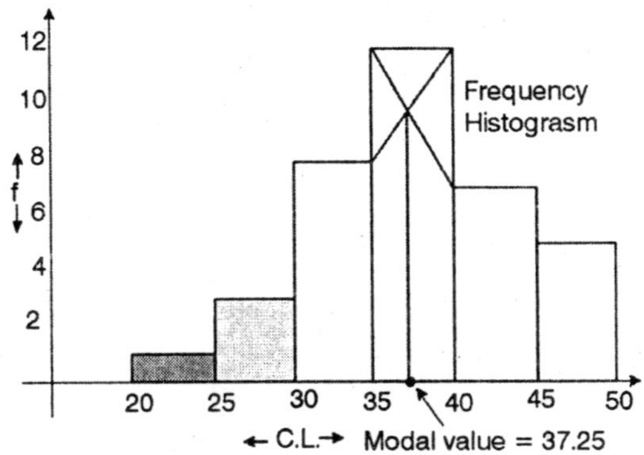

MERITS OF MODE

- It is simple to calculate.
- In individual or discrete distribution it can be located by mere inspection.
- It is easy to understand. Everyone is used to the idea of average size of a garment, an average American etc.
- It is not isolated like the median as it is the most common item.
- Like the Average mean, it is not a value which cannot be found in the series.
- It is not necessary to know all the items. What we need the point of maximum density frequency.
- It is not affected by sampling fluctuations.

DEMERITS

- It is ill defined.
- It is not based on all observations.
- It is not capable of further algebraic treatment.
- It is not a good representative of the data.

- Sometimes there are more than one values of mode.

EMPIRICAL RELATION BETWEEN MEAN, MEDIAN AND MODE

A distribution in which the values of mean, median and mode coincide (*i.e.* mean = median = mode) is known as a symmetrical distribution. Conversely, when values of mean, median and mode are not equal the distribution is known as asymmetrical or skewed distribution. In moderately skewed or asymmetrical distribution a very important relationship exists among these three measures of central tendency. In such distributions the distance between the mean and median is about one-third of the distance between the mean and mode, as will be clear from the diagrams 1 and 2. Karl Pearson expressed this relationship as:

$$\text{Mode} = \text{mean} - 3\,[\text{mean} - \text{median}]$$

$$\text{Mode} = 3\,\text{median} - 2\,\text{mean and Median}$$

$$= \text{mode} + \frac{2}{3}[\text{mean} - \text{mode}]$$

Knowing any two values, the third can be computed.

Example: Given median = 20.6, mode = 26 Find mean.

Solution:

$$\text{Mode} = 3\,\text{median} - 2\text{mean}$$

$$\text{Mean} = \frac{1}{2}[3\,\text{median} - \text{mode}]$$

$$\text{Mean} = \frac{1}{2}[3(20.6) - (26)]$$

$$\text{Mean} = \frac{1}{2}$$

$$\text{Mean} = 17.9$$

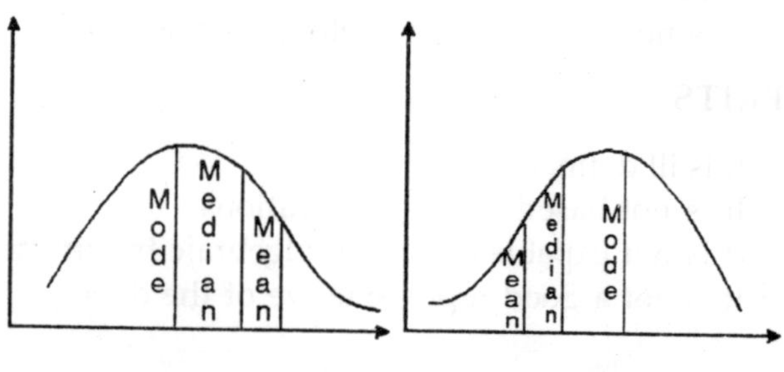

11

Sampling Theory

CONCEPT

Sampling allows statisticians to draw conclusions about a whole by examining a part. It enables us to estimate characteristics of a population by directly observing a portion of the entire population. Researchers are not interested in the sample itself, but in what can be learned from the survey—and how this information can be applied to the entire population.

It is essential that a sample survey be correctly defined and organized. If the wrong questions are posed to the wrong people, statisticians will not receive information that will be useful when applied to the entire population.

In the context of a national statistical agency like Statistics Canada, the following steps are needed to select a sample and ensure that this sample will fulfill its goals.

ESTABLISH THE SURVEY'S OBJECTIVES

The first step in planning a useful and efficient survey is to specify the objectives with as much detail as possible. Without objectives, the survey is unlikely to generate usable results. Clarifying the aims of the survey is critical to its ultimate success. The initial users and uses of the data should be identified at this stage. The pros and cons of a census versus a sample survey or the use of administrative records should be evaluated and a decision made as to the most appropriate method. (At this point, we will assume that a sample survey is the best way to proceed in order to obtain the information we need. This assumption will hold true for the remainder of the sample selection steps, even though many of the steps mentioned will also apply to the other methods.)

Define The Target Population

The target population is the total population for which the information is required. For example, if you were to conduct a survey about the most popular types of cars in Saskatchewan, then the target population would be every car in Saskatchewan. The units that make up the population must be described in terms of characteristics that clearly identify them.

Specifically, the target population is defined by the following characteristics:

- Nature of data required: about persons, hospitals, schools, etc.
- Geographic location: the geographic boundaries of the population have to be determined, as well as the level of geographic detail required for the survey estimate (by province, by city, etc.).
- Reference period: the time period covered by the survey.
- Other characteristics, such as socio-demographic characteristics (interest in a particular age group, for example) or type of industry.

Decide on the Data to be Collected

The data requirements of the survey must be established. To ensure that the requirements are operationally sound, the necessary data terms and definitions also need to be determined.

Set The Level of Precision

As mentioned in the section on Sampling error, there is a level of uncertainty associated with estimates coming from a sample. For example, if you are trying to estimate the average distance between home and school for students in your class of 25 from a sample of 5 persons, your estimate will depend on who the 5 sampled students are. If the 5 sampled students also live close to the school, the results will not be able to represent the class accurately. This sample-to-sample variation is what causes the sampling error. Statisticians can estimate the sampling error associated with a particular sampling plan, and try to minimize it.

When designing a survey, the acceptable level of uncertainty in the survey estimates has to be established. This level depends on what the end use of the results will be and on the size of the overall budget. The bigger the budget, the more resources available, and thus, less chance for error. And if the end result is to serve a specific purpose, then the acceptable level of uncertainty would be smaller than an end result that is simply looking for general trends.

The level of uncertainty will also be determined by the sample size. Increasing the sample size will decrease the sampling error. (If you sample 24 out of 25 students in your class, there will not be as much sample-to-sample variation as there would be if you only sampled 5 students from among the 25 possible samples.)

THE SAMPLE DESIGN

Once the objectives, guidelines and definitions have been worked out, the statistician can work on the survey plan.

The survey plan is divided into three parts:

- Sample design: how the sample will be collected.
- Estimation techniques: how the results from the sample will be extended to the whole population.
- Measures of precision: how the sampling error will be measured.

The estimation techniques and measures of precision are discussed in a later section. For the moment, we will look at the sample design.

The following steps lead to the complete determination of the sample design:

1. Determine what the survey population will be (*e.g.*, students, men aged 20 to 35, newborn babies, etc.).
2. Choose the most appropriate survey time frame.

3. Define the survey units.
4. Establish the sample size (*e.g.*, a sample of 100 from a population of 1,000).
5. Select a sampling method.

The Survey Population

The target population must be defined early in the survey-designing process. This is the population for which information is required. However, some members of the population have to be excluded because of operational constraints: the high cost of collecting data in some remote areas, the difficulty of identifying and contacting certain components of the target population, etc. For example, because it would be too difficult to locate and survey each car owned by every resident in Saskatchewan, a survey population of just the major cities and towns might be conducted instead. When some of the members of the target population are excluded, we call the included population the survey population or, what is sometimes called, an *observed population*. The target population is the population we want to observe while the survey population is the population we can observe.

The goal of this process is to have the survey population as close as possible to the target population. It is also very important that the users of the data be informed of the differences between the two populations, as the results of the survey will apply only to the survey population.

For example, a target population for a survey could be all Canadians aged 15 years and over (on a particular reference date), while the survey population could exclude residents of the Yukon, Nunavut and Northwest Territories, persons living on Aboriginal reserves, full-time members of the Canadian Armed Forces and residents of institutions. These Canadians might be excluded for various reasons: to survey people in the territories might prove to be difficult and expensive, military personnel may not be available for surveying if they are out on a mission, etc. Using this example, about 2% of the target population would be excluded from the survey population.

The Survey Frame

The survey frame, also called the sampling frame, is the tool used to gain access to the population. There are two types of frames: list frames and area frames. A list frame is just a list of names and addresses that provide direct access to 'individuals' (*e.g.*, a list of hospitals, a list of restaurants, a list of students at a university). Area frames are a list of geographic areas that provide indirect access to individuals (*e.g.*, the neighbourhoods in a city). This type of access is called indirect because first, a list of geographic areas must be selected and then, access to individuals within each selected area must be worked out.

For instance, suppose that you were surveying a rural town in Quebec to see what percentage of residents are farmers. If you were provided with an area frame, then you would be able to locate which roads to visit, but you would still have to find out the names and addresses of the residents on each road. When there is no single frame that is appropriate, multiple frames can be used. Some sampling techniques using both types of frames will be discussed later.

A good frame should be complete and up-to-date; no member of the survey population should be excluded from the frame or duplicated on the frame (represented more than once); and no unit that is not part of the population (*e.g.*, deceased persons) should be on the frame. The frame chosen will impact the selected survey population. For instance, if a list of telephone numbers is used to select a sample of households, then all households without telephones are excluded from the survey population.

The survey units

There are three types of units that have to be accurately identified in order to avoid problems during the selection, data collection and data analysis stages. They are as follows:

- The sampling unit is part of the frame and therefore subject to being selected.
- The respondent unit or reporting unit provides the information needed by the survey.

- The unit of reference or unit of analysis—the unit about which information is provided—is used to analyse the survey results.

For example, in a survey about newborns in Edmonton, the sampling unit might be a household, the reporting unit one of the parents or a legal guardian, and the unit of reference the baby. The sampling units may differ depending on the frame used. This is why the survey population, survey frame and survey units are defined in conjunction with one another.

The Sample Size

The level of precision needed for the survey estimates will impact the sample size. However, it is not as easy to determine the sample size as one may think. Generally, the actual sample size of a survey is a compromise between the level of precision to be achieved, the survey budget and any other operational constraints, such as budget and time. In order to achieve a certain level of precision, the sample size will depend, among other things, on the following factors:

- *The variability of the characteristics being observed*: If every person in a population had the same salary, then a sample of one person would be all you would need to estimate the average salary of the population. If the salaries are very different, then you would need a bigger sample in order to produce a reliable estimate.
- *The population size*: To a certain extent, the bigger the population, the bigger the sample needed. But once you reach a certain level, an increase in population no longer affects the sample size. For instance, the necessary sample size to achieve a certain level of precision will be about the same for a population of one million as for a population twice that size.
- *The sampling and estimation methods*: Not all sampling and estimation methods have the same level of efficiency. You will need a bigger sample if your method is not the most efficient. But because of operational constraints and the unavailability of an

adequate frame, you cannot always use the most efficient technique.

The Sampling Method

There are two types of sampling methods: probability sampling and non-probability sampling. The difference between them is that in probability sampling, every unit has a 'chance' of being selected, and that chance can be quantified. This is not true for non-probability sampling; every item in a population does not have an equal chance of being selected. The next section will describe features of both types of sampling and detail some of the methods related to each type.

PROBABILITY SAMPLING

Probability sampling involves the selection of a sample from a population, based on the principle of randomization or chance. Probability sampling is more complex, more time-consuming and usually more costly than non-probability sampling. However, because units from the population are randomly selected and each unit's probability of inclusion can be calculated, reliable estimates can be produced along with estimates of the sampling error, and inferences can be made about the population.

There are several different ways in which a probability sample can be selected. The method chosen depends on a number of factors, such as the available sampling frame, how spread out the population is, how costly it is to survey members of the population and how users will analyse the data. When choosing a probability sample design, your goal should be to minimize the sampling error of the estimates for the most important survey variables, while simultaneously minimizing the time and cost of conducting the survey.

Simple Random Sampling

In *simple random sampling,* each member of a population has an equal chance of being included in the sample. Also, each combination of members of the population has an equal chance of composing the sample. Those two properties are what defines simple random sampling. To select a simple random sample, you need to list all of the units in the survey population.

Example: To draw a simple random sample from a telephone book, each entry would need to be numbered sequentially. If there were 10,000 entries in the telephone book and if the sample size were 2,000, then 2,000 numbers between 1 and 10,000 would need to be randomly generated by a computer. Each number will have the same chance of being generated by the computer (in order to fill the simple random sampling requirement of an equal chance for every unit). The 2,000 telephone entries corresponding to the 2,000 computer-generated random numbers would make up the sample.

Simple random sampling can be done with or without replacement. A sample with replacement means that there is a possibility that the sampled telephone entry may be selected twice or more. Usually, the simple random sampling approach is conducted without replacement because it is more convenient and gives more precise results. For the purpose of these descriptions, when we discuss simple random sampling, we will refer to sampling without replacement.

Simple random sampling is the easiest method of sampling and it is the most commonly used. Advantages of this technique are that it does not require any additional information on the frame (such as geographic areas) other than the complete list of members of the survey population along with information for contact. Also, since simple random sampling is a simple method and the theory behind it is well established, standard formulas exist to determine the sample size, the estimates and so on, and these formulas are easy to use.

On the other hand, this technique makes no use of auxiliary information present on the frame (*i.e.*, number of employees in each business) that could make the design of the sample more efficient. And although it is easy to apply simple random sampling to small populations, it can be expensive and unfeasible for large populations because all elements must be identified and labeled prior to sampling. It can also be expensive if personal interviewers are required since the sample may be geographically spread out across the population.

A lottery draw is a good example of simple random sampling. For example, when a sample of 6 numbers is randomly generated from a population of 49, each number has an equal chance of being selected and each combination of 6 numbers has the same chance of being the winning combination. Even though people tend to avoid combinations such as 1-2-3-4-5-6, it has the same chance of being the winning set of numbers as the combination of 8-15-21-28-32-40.

Example: Suppose your school has 500 students and you need to conduct a short survey on the quality of the food served in the cafeteria. You decide that a sample of 10 students should be sufficient for your purposes. In order to get your sample, you assign a number from 1 to 500 to each student in your school. To select the sample, you use a table of randomly generated numbers. All you have to do is pick a starting point in the table (a row and column number) and look at the random numbers that appear there. In this case, since the data run into three digits, the random numbers would need to contain three digits as well. Ignore all random numbers after 500 because they do not correspond to any of the students in the school. Remember that the sample is without replacement, so if a number recurs, skip over it and use the next random number. The first 10 different numbers between 001 and 500 make up your sample.

Example: Imagine that you own a movie theatre and you are offering a special horror movie film festival next month. To decide which horror movies to show, you survey moviegoers asking them which of the listed movies are their favourites. To create the list of movies needed for your survey, you decide to sample 100 of the 1,000 best horror movies of all time. The horror movie population is divided evenly into classic movies (those filmed in or before 1969) and modern movies (those filmed in or later than 1970). One way of getting a sample would be to write out all of the movie titles on slips of paper and place them in an empty box. Then, draw out 100 titles and you will have your sample. By using this approach, you will have ensured that each movie had an equal chance of selection. You can also calculate the probability of a given movie being selected. Since we know the sample size (n) and the total population (N), calculating the probability of being included in the sample becomes a simple matter of division:

Probability of being selected (same for each horror movie)
= (n ÷ N) × 100%
= (100 ÷ 1,000) × 100%
= 10%

This means that every movie title on the list has a 10% or a 1 in 10 chance of being selected.

You can see that that one disadvantage of simple random sampling (not the only disadvantage, but an important one) is that even if you know that the population is made up of 500 classic movies and 500 modern movies and you know each movie's release date from the sampling frame, no use is made of this information. This sample might contain 77 classic movies and 23 modern movies, which would not be representative of the whole horror movie population.

There are ways to overcome this problem (these will be briefly discussed in the Estimation section), but there are also ways to account for this information. (This will also be discussed later, under the section on Stratified sampling.)

Systematic Sampling

Sometimes called *interval sampling, systematic sampling* means that there is a gap, or interval, between each selected unit in the sample.

In order to select a systematic sample, you need to follow these steps:

1. Number the units on your frame from 1 to N (where N is the total population size).
2. Determine the sampling interval (K) by dividing the number of units in the population by the desired sample size. For example, to select a sample of 100 from a population of 400, you would need a sampling interval of 400 ÷ 100 = 4. Therefore, K = 4.

You will need to select one unit out of every four units to end up with a total of 100 units in your sample.

3. Select a number between one and K at random. This number is called *the random start* and would be the first number included in your sample. Using the sample above, you would select a number between 1 and 4 from a table of random numbers. If you choose 3, the third unit on your frame would be the first unit included in your sample; if you choose 2, your sample would start with the second unit on your frame.
4. Select every K^{th} (in this case, every fourth) unit after that first number. For example, the sample might consist of the following units to make up a sample of 100: 3 (the random start), 7, 11, 15, 19...395, 399 (up to N, which is 400 in this case).

Using the example above, you can see that with a systematic sample approach there are only four possible samples that can be selected, corresponding to the four possible random starts:

- 1, 5, 9, 13...393, 397
- 2, 6, 10, 14...394, 398
- 3, 7, 11, 15...395, 399
- 4, 8, 12, 16...396, 400

Each member of the population belongs to only one of the four samples and each sample has the same chance of being selected. From that, we can see that each unit has a one in four chance of being selected in the sample. This is the same probability as if a simple random sampling of 100 units was selected.

The main difference is that with simple random sampling, any combination of 100 units would have a chance of making up the sample, while with systematic sampling, there are only four possible samples. From that, we can see how precise systematic sampling is compared with simple random sampling.

The population's order on the frame will determine the

possible samples for systematic sampling. If the population is randomly distributed on the frame, then systematic sampling should yield results that are similar to simple random sampling.

This method is often used in industry, where an item is selected for testing from a production line to ensure that machines and equipment are of a standard quality. For example, a tester in a manufacturing plant might perform a quality check on every 20th product in an assembly line. The tester might choose a random start between the numbers 1 and 20. This will determine the first product to be tested; every 20th product will be tested thereafter.

Interviewers can use this sampling technique when questioning people for a sample survey. The market researcher might select, for example, every 10th person who enters a particular store, after selecting the first person at random. The surveyor may interview the occupants of every fifth house on a street, after randomly selecting one of the first five houses.

Example: Imagine you have to conduct a survey on student housing for your university or college. Your school has an enrolment of 10,000 students and you want to take a systematic sample of 500 students. In order to do this, you must first determine what your sampling interval (K) would be:

Total population ÷ sample size = sampling interval

$N \div n = K$

$= 10{,}000 \div 500$

$= 20$

To begin this systematic sample, all students would have to be assigned sequential numbers. The starting point would be chosen by selecting a random number between 1 and 20. If this number were 9, then the 9th student on the list would be selected along with every 20th student thereafter. The sample of students would be those corresponding to student numbers 9, 29, 49, 69...9,929, 9,949, 9,969 and 9,989.

In the examples used thus far, the sampling interval K was always a whole number, but this is not always the case. For example, if you want a sample of 30 from a population of

740, your sampling interval (or K) will be 24.7. In these cases, there are a few options to make the number easier to work with. You can round the number—either round it up to the nearest whole number or round it down. Rounding down will ensure that you select at least the number of units you originally wanted (and you can then delete some units to get the exact sample size you wanted). Techniques exist to adapt systematic sampling to the case where N (total population) is not a multiple of n (sample size), but still give a sample exactly the same as the n units. These techniques will not be discussed here.

The advantages of systematic sampling are that the sample selection cannot be easier (you only get one random number—the random start—and the rest of the sample automatically follows) and that the sample is distributed evenly over the listed population.

The biggest drawback of the systematic sampling method is that if there is some cycle in the way the population is arranged on a list and if that cycle coincides in some way with the sampling interval, the possible samples may not be representative of the population.

Example: Suppose you run a large grocery store and have a list of the employees in each section. The grocery store is divided into the following 10 sections: deli counter, bakery, cashiers, stock, meat counter, produce, pharmacy, photo shop, flower shop and dry cleaning.

Each section has 10 employees, including a manager (making 100 employees in total). Your list is ordered by section, with the manager listed first and then, the other employees by descending order of seniority.

If you wanted to survey your employees about their thoughts on their work environment, you might choose a small sample to answer your questions. If you use a systematic sampling approach and your sampling interval is 10, then you could end up selecting only managers or the newest employees in each section. This type of sample would not give you a complete or appropriate picture of your employees' thoughts.

Sampling with Probability Proportional to Size

Probability sampling requires that each member of the survey population have a chance of being included in the sample, but it does not require that this chance be the same for everyone. If there is information available on the frame about the size of each unit (*e.g.*, number of employees for each business) and if those units vary in size, this information can be used in the sampling selection in order to increase the efficiency. This is known as *sampling with probability proportional to size* (PPS). With this method, the bigger the size of the unit, the higher the chance it has of being included in the sample. For this method to bring increased efficiency, the measure of size needs to be accurate. This is a more complex sampling method that will not be discussed in further detail here.

Stratified Sampling

Using *stratified sampling*, the population is divided into homogeneous, mutually exclusive groups called strata, and then independent samples are selected from each stratum. Any of the sampling methods mentioned in this section (and others that exist) can be used to sample within each stratum. The sampling method can vary from one stratum to another. When simple random sampling is used to select the sample within each stratum, the sample design is called *stratified simple random sampling*. A population can be stratified by any variable that is available for all units on the sampling frame prior to sampling (*e.g.*, age, sex, province of residence, income, etc.).

Why do we need to create strata? There are many reasons, the main one being that it can make the sampling strategy more efficient. It was mentioned earlier that you need a larger sample to get a more accurate estimation of a characteristic that varies greatly from one unit to the other than for a characteristic that does not. For example, if every person in a population had the same salary, then a sample of one individual would be enough to get a precise estimate of the average salary.

This is the idea behind the efficiency gain obtained with stratification. If you create strata within which units share similar characteristics (*e.g.*, income) and are considerably different from units in other strata (*e.g.*, occupation, type of dwelling) then you would only need a small sample from each stratum to get a precise estimate of total income for that stratum. Then you could combine these estimates to get a precise estimate of total income for the whole population. If you were to use a simple random sampling approach in the whole population without stratification, the sample would need to be larger than the total of all stratum samples to get an estimate of total income with the same level of precision.

Stratified sampling ensures an adequate sample size for sub-groups in the population of interest. When a population is stratified, each stratum becomes an independent population and you will need to decide the sample size for each stratum.

Example: Suppose you want to estimate how many high school students have part-time jobs at the national level and also in each province. If you were to select a simple random sample of 25,000 people from a list of all high school students in Canada (assuming such a list was available for selection), you would end up on average with just a little over 100 people from Prince Edward Island, since they account for less than half of a per cent of the whole Canadian population. This sample would probably not be large enough for the kind of detailed analysis you had in mind. Stratifying your list by province, again assuming that this information is available, and then selecting a sample size for each province would allow you to decide on the exact sample size needed for that specific province. Thus, in order to get good representation of Prince Edward Island, you would use a larger sample than the one allotted to it by the simple random sampling approach.

Example: An Ontario school board wanted to assess student opinion on dropping Grade 13 from the secondary school programme. They decided to survey students from Elmsview High School. To ensure a representative sample of students from all grade levels, the school board used a stratified sampling technique.

In this case, the strata were the five grade levels (grades 9 to 13). The school board then selected a sample within each stratum. The students selected in this sample were extracted using simple random or systematic sampling, making up a total sample of 100 students.

Stratification is most useful when the stratifying variables are:

- Simple to work with,
- Easy to observe, and
- Closely related to the topic of the survey.

Cluster Sampling

Sometimes it is too expensive to spread a sample across the population as a whole. Travel costs can become expensive if interviewers have to survey people from one end of the country to the other. To reduce costs, statisticians may choose a *cluster sampling* technique.

Cluster sampling divides the population into groups or clusters. A number of clusters are selected randomly to represent the total population, and then all units within selected clusters are included in the sample. No units from non-selected clusters are included in the sample—they are represented by those from selected clusters. This differs from stratified sampling, where some units are selected from each group. Examples of clusters are factories, schools and geographic areas such as electoral subdivisions. The selected clusters are used to represent the population.

Example: Suppose you are a representative from an athletic organization wishing to find out which sports Grade 11 students are participating in across Canada. It would be too costly and lengthy to survey every Canadian in Grade 11, or even a couple of students from every Grade 11 class in Canada. Instead, 100 schools are randomly selected from all over Canada.

These schools provide clusters of samples. Then every Grade 11 student in all 100 clusters is surveyed. In effect, the students in these clusters represent all Grade 11 students in Canada.

Example: Imagine that the municipal council of a small city wants to investigate the use of health care services by residents. First, the council requests from Statistics Canada electoral subdivision maps that identify and label each city block. From these maps, the council creates a list of all city blocks. This list will serve as the sampling frame.

Every household in that city belongs to a city block, and each city block represents a cluster of households. The council randomly picks a number of city blocks. Using the simple random sample approach, then the council creates a list of all households in the selected city blocks; these households make up the survey sample.

As mentioned, cost reduction is a reason for using cluster sampling. It creates 'pockets' of sampled units instead of spreading the sample over the whole territory. Another reason is that sometimes a list of all units in the population (a requirement when conducting simple random sample, systematic sample or sampling with probability proportional to size) is not available, while a list of all clusters is either available or easy to create.

In most cases, the main drawback is a loss of efficiency when compared with simple random sampling. It is usually better to survey a large number of small clusters instead of a small number of large clusters. This is because neighbouring units tend to be more alike, resulting in a sample that does not represent the whole spectrum of opinions or situations present in the overall population. In the two previous examples, students in the same school tend to participate in the same types of sports (depending on the facilities available at their school); similarly, elderly people have a tendency to live in specific neighbourhoods and to be heavy users of health services. Another drawback to cluster sampling is that you do not have total control over the final sample size. Since not all schools have the same number of Grade 11 students and city blocks do not all have the same number of households, and you must interview every student or household in your sample, the final size may be larger or smaller than you expected.

Multi-stage Sampling

Multi-stage sampling is like the cluster method, except that it involves picking a sample from within each chosen cluster, rather than including all units in the cluster. This type of sampling requires at least two stages. In the first stage, large groups or clusters are identified and selected. These clusters contain more population units than are needed for the final sample.

In the second stage, population units are picked from within the selected clusters (using any of the possible probability sampling methods) for a final sample. If more than two stages are used, the process of choosing population units within clusters continues until there is a final sample.

Example: In Example, a cluster sample would choose 100 schools and then interview every Grade 11 student from those schools. Instead in multi-stage sampling, you could select more schools, get a list of all Grade 11 students from these selected schools and select a random sample (*e.g.*, simple random sample) of students from each school. This would be a two-stage sampling design.

You could also get a list of all Grade 11 classes in the selected schools, pick a random sample of classes from each of those schools, get a list of all the students in the selected classes and finally select a random sample of students from each class. This would be a three-stage sampling design. Each time we add a stage, the process becomes more complex.

Now imagine that each school has on average 80 Grade 11 students. Cluster sampling would then give your organization a sample of about 8,000 students (100 schools x 80 Grade 11 students). If you wanted a bigger sample, you could select schools with more students; and for a smaller sample you could select schools with fewer students.

One way to control the sample size would be to stratify the schools into large, medium and small sizes (in terms of the number of Grade 11 students) and select a sample of schools from each stratum. This is called *stratified cluster sampling.*

With a three-stage design, you could select a sample of 400 schools, then select two Grade 11 classes per school (assuming that there are two or more Grade 11 classes per school). Finally, you could select 10 students per class. This way, you still end up with a sample of about 8,000 students

(400 schools x 2 classes x 10 students), but the sample will be more spread out.

You can see from this example that with multi-stage sampling, you still have the benefit of a more concentrated sample for cost reduction. However, the sample is not as concentrated as other clusters and the sample size is still bigger than for a simple random sample size.

Also, you do not need to have a list of all of the students in the population. All you need is a list of the classes from the 400 schools and a list of the students from the 800 classes. Admittedly, more information is needed in this type of sample than what is required in cluster sampling. However, multi-stage sampling still saves a great amount of time and effort by not having to create a list of all of the units in a population.

Multi-phase Sampling

A *multi-phase sample* collects basic information from a large sample of units and then, for a subsample of these units, collects more detailed information. The most common form of multi-phase sampling is two-phase sampling (or double sampling), but three or more phases are also possible.

Multi-phase sampling is quite different from multi-stage sampling, despite the similarities in name. Although multi-phase sampling also involves taking two or more samples, all samples are drawn from the same frame and at each phase the units are structurally the same. However, as with multi-stage sampling, the more phases used, the more complex the sample design and estimation will become.

Multi-phase sampling is useful when the frame lacks auxiliary information that could be used to stratify the population or to screen out part of the population.

Example: Suppose that an organization needs information about cattle farmers in Alberta, but the survey frame lists all types of farms—cattle, dairy, grain, hog, poultry and produce. To complicate matters, the survey frame does not provide any auxiliary information for the farms listed there.

A simple survey could be conducted whose only question is "Is part or all of your farm devoted to cattle farming?" With only one question, this survey should have a low cost per interview (especially if done by telephone) and, consequently, the organization should be able to draw a large sample. Once the first sample has been drawn, a second, smaller sample can be extracted from among the cattle farmers and more detailed questions asked of these farmers. Using this method, the organization avoids the expense of surveying units that are not in this specific scope (*i.e.*, non-cattle farmers).

Multi-phase sampling can be used when there is insufficient budget to collect information from the whole sample, or when doing so would create excessive burden on the respondent, or even when there are very different costs of collection for different questions on a survey.

Example: A health survey asks participants some basic questions about their diet, smoking habits, exercise routines and alcohol consumption. In addition, the survey requires that respondents subject themselves to some direct physical tests, such as running on a treadmill or having their blood pressure and cholesterol levels measured.

Filling out questionnaires or interviewing participants are relatively inexpensive procedures, but the medical tests require the supervision and assistance of a trained health practitioner, as well as the use of an equipped laboratory, both of which can be quite costly. The best way to conduct this survey would be to use a two-phase sample approach. In the first phase, the interviews are performed on an appropriately sized sample. From this sample, a smaller sample is drawn. This second sample will take part in the medical tests.

NON-PROBABILITY SAMPLING

The difference between *probability* and *non-probability sampling* has to do with a basic assumption about the nature of the population under study. In probability sampling, every item has a chance of being selected. In non-probability sampling, there is an assumption that there is an even distribution of characteristics within the population. This is what makes the researcher believe that any sample would be representative and because of that, results will be accurate. For probability sampling, randomization is a feature of the selection process, rather than an assumption about the structure of the population.

In non-probability sampling, since elements are chosen arbitrarily, there is no way to estimate the probability of any one element being included in the sample. Also, no assurance is given that each item has a chance of being included, making it impossible either to estimate sampling variability or to identify possible bias.

Reliability cannot be measured in non-probability sampling; the only way to address data quality is to compare some of the survey results with available information about the population. Still, there is no assurance that the estimates will meet an acceptable level of error. Statisticians are reluctant to use these methods because there is no way to measure the precision of the resulting sample.

Despite these drawbacks, non-probability sampling methods can be useful when descriptive comments about the sample itself are desired. Secondly, they are quick, inexpensive and convenient. There are also other circumstances, such as in applied social research, when it is unfeasible or impractical to conduct probability sampling. Statistics Canada uses probability sampling for almost all of its surveys, but uses non-probability sampling for questionnaire testing and some preliminary studies during the development stage of a survey.

Most non-sampling methods require some effort and organization to complete, but others, like convenience sampling, are done casually and do not need a formal plan of action.

Convenience or Haphazard Sampling

Convenience sampling is sometimes referred to as *haphazard* or *accidental sampling*. It is not normally representative of the target population because sample units are only selected if they can be accessed easily and conveniently.

There are times when the average person uses convenience sampling. A food critic, for example, may try several appetizers or entrees to judge the quality and variety of a menu. And television reporters often seek so-called 'people-on-the-street interviews' to find out how people view an issue. In both these examples, the sample is chosen randomly, without use of a specific survey method. The obvious advantage is that the method is easy to use, but that advantage is greatly offset by the presence of bias. Although useful applications of the technique are limited, it can deliver accurate results when the population is homogeneous.

For example, a scientist could use this method to determine whether a lake is polluted. Assuming that the lake water is well-mixed, any sample would yield similar information. A scientist could safely draw water anywhere on the lake without fretting about whether or not the sample is representative.

Examples of convenience sampling include:

- The female moviegoers sitting in the first row of a movie theatre
- The first 100 customers to enter a department store
- The first three callers in a radio contest.

Volunteer Sampling

As the term implies, this type of sampling occurs when people volunteer their services for the study. In psychological experiments or pharmaceutical trials (drug testing), for example, it would be difficult and unethical to enlist random participants from the general public. In these instances, the sample is taken from a group of volunteers. Sometimes, the researcher offers payment to entice respondents. In exchange, the volunteers accept the possibility of a lengthy, demanding or sometimes unpleasant process.

Sampling voluntary participants as opposed to the general population may introduce strong biases. Often in opinion polling, only the people who care strongly enough about the subject one way or another tend to respond. The silent majority does not typically respond, resulting in large selection bias. Television and radio media often use call-in polls to informally query an audience on their views. The Much Music television channel uses this kind of survey in their CombatZone programme. The programme asks viewers to cast a vote for one of two music videos by telephone, e-mail or through their online website.

Oftentimes, there is no limit imposed on the frequency or number of calls one respondent can make. So, unfortunately, a person might be able to vote repeatedly. It should also be noted that the people who contribute to these surveys might have different views than those who do not.

Judgement Sampling

This approach is used when a sample is taken based on certain judgements about the overall population. The underlying assumption is that the investigator will select units that are characteristic of the population. The critical issue here is objectivity: how much can judgment be relied upon to arrive at a typical sample? *Judgement sampling* is subject to the researcher's biases and is perhaps even more biased than haphazard sampling. Since any preconceptions the researcher may have are reflected in the sample, large biases can be introduced if these preconceptions are inaccurate.

Statisticians often use this method in exploratory studies like pre-testing of questionnaires and focus groups. They also prefer to use this method in laboratory settings where the choice of experimental subjects (*i.e.*, animal, human, vegetable) reflects the investigator's pre-existing beliefs about the population.

One advantage of judgement sampling is the reduced cost and time involved in acquiring the sample.

Quota Sampling

This is one of the most common forms of non-probability sampling. Sampling is done until a specific number of units (quotas) for various sub-populations have been selected. Since there are no rules as to how these quotas are to be filled, *quota sampling* is really a means for satisfying sample size objectives for certain sub-populations. The quotas may be based on population proportions. For example, if there are 100 men and 100 women in a population and a sample of 20 are to be drawn to participate in a cola taste challenge, you may want to divide the sample evenly between the sexes—10 men and 10 women. Quota sampling can be considered preferable to other forms of non-probability sampling (*e.g.*, judgement sampling) because it forces the inclusion of members of different sub-populations.

Quota sampling is somewhat similar to stratified sampling in that similar units are grouped together. However, it differs in how the units are selected. In probability sampling, the units are selected randomly while in quota sampling it is usually left up to the interviewer to decide who is sampled. This results in selection bias. Thus, quota sampling is often used by market researchers (particularly for telephone surveys) instead of stratified sampling, because it is relatively inexpensive and easy to administer and has the desirable property of satisfying population proportions. However, it disguises potentially significant bias. As with all other non-probability sampling methods, in order to make inferences about the population, it is necessary to assume that persons selected are similar to those not selected. Such strong assumptions are rarely valid.

Example: The student council at Cedar Valley Public School wants to gauge student opinion on the quality of their extracurricular activities. They decide to survey 100 of 1,000 students using the grade levels (7 to 12) as the sub-population. The table below gives the number of students in each grade level.

Table. Number of students enrolled at Cedar Valley Public School, by grade

Grade level of students sample	Number of students (%)	Percentage	Quota of studentsin of 100
7	150	15	15
8	220	22	22
9	160	16	16
10	150	15	15

11	200	20	20
12	120	12	12
Total	**1,000**	**100**	**100**

The student council wants to make sure that the percentage of students in each grade level is reflected in the sample.

The formula is:

Percentage of students in Grade 10

= (number of students ÷ number of students) x 100%

= (150 ÷ 1,000) × 100

= 15%

Since 15% of the school population is in Grade 10, 15% of the sample should contain Grade 10 students. Therefore, use the following formula to calculate the number of Grade 10 students that should be included in the sample:

Sample of Grade 10 students

= (15% of 100) × 100

= 0.15 × 100

= 15 students

The main difference between stratified sampling and quota sampling is that stratified sampling would select the students using a probability sampling method such as simple random sampling or systematic sampling. In quota sampling, no such technique is used. The 15 students might be selected by choosing the first 15 Grade 10 students to enter school on a certain day, or by choosing 15 students from the first two rows of a particular classroom. Keep in mind that those students who arrive late or sit at the back of the class may hold different opinions from those who arrived earlier or sat in the front.

The main argument against quota sampling is that it does not meet the basic requirement of randomness. Some units may have no chance of selection or the chance of selection may be unknown. Therefore, the sample may be biased.

It is common, but not necessary, for quota samples to use random selection procedures at the beginning stages, much in the same way as probability sampling does. For instance, the first step in multi-stage sampling would be randomly selecting the geographic areas. The difference is in the selection of the units in the final stages of the process.

In multi-stage sampling, units are based on up-to-date lists for selected areas and a sample is selected according to a random process. In quota sampling, by contrast, each interviewer is instructed on how many of the respondents should be men and how many should be women, as well as how many people should represent the various age groups. The quotas are therefore calculated from available data for the population, so that the sexes, age groups or other demographic variables are represented in the correct proportions. But within each quota, interviewers may fail to secure a representative sample of respondents.

For example, suppose that an organization wants to find out information about the occupations of men aged 20 to 25. An interviewer goes to a university campus and selects the first 50 men aged 20 to 25 that she comes across and who agree to participate in her organization's survey. However, this sample does not mean that these 50 men are representative of all men aged 20 to 25.

Quota sampling is generally less expensive than random sampling. It is also easy to administer, especially considering the tasks of listing the whole population, randomly selecting the sample and following-up on non-respondents can be omitted from the procedure. Quota sampling is an effective sampling method when information is urgently required and can be carried out independent of existing sampling frames. In many cases where the population has no suitable frame, quota sampling may be the only appropriate sampling method.

ESTIMATION

As we now know, the goal of conducting surveys is to obtain information about a particular population. When the sample has been selected and the information collected and processed, there still remains the task of linking the information gathered from the sample back to the overall population.

Estimation is the process of determining a likely value for a variable in the survey population, based on information collected from the sample. Researchers are usually interested in looking at estimates of many statistics—totals, averages and proportions being the most frequent—for different variables.

For example, a sample survey could be used to produce any of the following statistics: estimates for the proportion of smokers among all people aged 15 to 24 in the population; the average earnings of men and women with a university degree; or the total number of cars possessed by the whole survey population.

Underpinning the estimation process is the sampling weight of a unit, which indicates the number of units in the population (including the sampling weight) that are represented by this sampled unit. The sampling weight is the inverse of the unit's probability of selection.

Example: Suppose that the city of Winnipeg has decided to award bus travellers with free one-year bus passes as a way of promoting its services. A simple random sample of 10 people is selected from the 30 passengers on a city bus. Since simple random sampling gives equal probability of selection to every member of the population (in this case, all passengers on the bus), each passenger had one chance out of three of being selected.

This translates into a sampling weight of three for every selected unit. This means that each person in the sample represents three persons in the population—himself or herself, plus two other persons

To estimate this sampling weight, one could take the survey information for the 10 selected passengers and copy it three times to create an artificial population of 30. Totals, averages or proportions for the real population could then be estimated by the corresponding statistics computed using the artificial population. Instead of doing this, however, survey statisticians attach a sampling weight to each unit in the sample and take this weight into account when estimating.

If one person in a sample (with a sampling weight of 18) had blue eyes and brown hair, then it is as if a total of 18 people in the population had blue eyes and brown hair.

Example: You are conducting a survey to determine the total number of people living on your street and the average number of cars owned by each household. You decide to select a systematic sample of 5 households from the 20 on your street and intend to use that sample to estimate the totals you are looking for.

The following table summarizes the information that you gathered during your interviews with the sampled households:

Table. Sample of households on Redwood Street

Household number	Number of persons	Number of cars	Probability of selection	Sampling weight
1	1	0	1/4	4
2	4	2	1/4	4
3	2	1	1/4	4
4	2	1	1/4	4
5	3	2	1/4	4

The selection probability of 1 in 4 comes from the fact that systematic sampling gives an equal chance of being selected to each household on your street. The sampling weight of 4 is just the inverse of that probability. When estimating, you have to look at the characteristics of each sampled household. In this case, it is decided that 4 households from the population of 20 on your street have the same characteristics.

In order to estimate the total number of persons living on your street, you have to multiply the number of persons in a household by the number of households in that sampling weight, then add up all the final numbers. For example, there are 4 one-person households (represented by Household number 1), 4 four-person households, 8 two-person households (four households represented by Household number 3 and four households represented by Household number 4) and 4 three-person households.

The estimation of the total number of persons would then be:

Estimated number of persons living on your street

$= (4 \times 1) + (4 \times 4) + (8 \times 2) + (4 \times 3)$

= 48 people

To estimate the average number of cars per household, you proceed in the same manner. Get an estimate of the total number of cars owned by households on your street and then, divide the estimate by the actual number of households on the street. For example, there are 4 households without a car (represented by Household number 1), 8 households with two cars (represented by Household number 2 and Household number 5), 8 households with one car each (represented by Household number 3 and Household number 4).

Estimated number of cars:

$= (4 \times 0) + (8 \times 2) + (8 \times 1)$

= 24 cars

Estimated average:

$= 24 \div 20$

= 1.2 cars per household

Self-weighting Designs

It is not always the case that all sampled units had the same sampling weight. Some designs give unequal probability of selection to units, resulting in units within the same sample having different sampling weights. Answers from one household or business could represent the answers for 200 units of the population, while the answers from another could represent only 50 units in the population.

When every unit in the sample has the same sampling weight, the sampling design is said to be self-weighted. This kind of design is time-saving and operationally convenient, particularly for large samples. Because every unit has the same weight, those weights can be ignored when estimating averages and proportions. The average for the sample gives an appropriate estimate of the average for the whole population.

Simple random sampling and systematic sampling are examples of self-weighted designs. In that sense, calculations could have been made easier in example. For instance, to estimate the average number of cars per household in the population, we could have used the same average as the one used in the sample. The 5 sampled households own a total of 6 cars, an average of 1.2 cars per household. This is the same result as that obtained using the sampling weight procedure.

Adjusting The Weights

Sometimes, the sampling weights are adjusted prior to estimation.

There are basically two reasons for weight adjustment:

- *Adjusting for non-response*: Using sampling weights for estimation works fine when you have been able to interview all selected units. In Example 2, if two of the five sampled households refused to answer or were unavailable at the time of the survey, you would only have answers for three households, thus representing only 12 of the 20 households on the street. The two non-responding units represent four households each. This means that we have no information on the number of persons or cars for 8 households on your street. In order to adjust for that, survey statisticians usually increase the weights of responding units to account for the loss of representativeness caused by non-response. The goal would be to use only the 3 units for which we have information, but still represent the 20 households on the street.
- *Adjusting for external information*: Sometimes, we know the actual total for one or more variables measured in the sample. In Example 3 of the Probability sampling section, a population of the 1,000 best horror movies was equally divided into 500 classic movies and 500 modern movies. Even though you knew this prior to sampling, you decided to select a simple random sample of 100 movies and ended up with 77 classic movies and 23 modern movies. Each of these movies has a weight of 10 (because you

selected 1 movie out of every 10 titles). Using the answers from the survey and the sampling weight, your sample would represent a population of 770 classic movies and 230 modern movies. This could lead to inaccurate estimates. One solution would be to decrease the weight of every sampled classic and increase the weight of every sampled modern movie so that your sample gives an estimate of 500 classics and 500 modern films in the population. This should reduce the distortion caused by a 'bad' sample.

Of course, stratifying by release date prior to sampling would have solved this problem. However, in a lot of cases, we have totals at the population level, but we don't know the attribute of each unit on the sampling frame. For example, from the Census of Population, we know how many men and women there are in a specific city, but all we have for sampling is a list of households. Thus, stratifying our population by sex would not be possible. Demographic projections by age and sex for each province are often used in social surveys to adjust sampling weights.

The weights adjusted for non-response and/or external counts are used for estimation, in the same way as the sampling weight was used in Example 1.

Other Estimation Methods

Using the weights to inflate the sample results is not the only estimation method that exists, but it is the simplest one and the only one that we will cover. Nevertheless, it is important to know that there exist some other methods that could lead to more precise estimates (*e.g.*, using auxiliary information). The estimation process has to take into account the sampling design that was used. Otherwise, the resulting estimates could be severely biased.

Estimating The Sampling Error

As mentioned before, any estimates derived from samples are subject to what is called the sampling error. This comes from the fact that only a part of the population was observed, instead of the whole. A different sample could have come up with different results.

The amount of variation that exists among the estimates from the different possible samples is what makes the sampling error. Of course, this sampling error is unknown, since we would need to know the answer for each unit of the population in order to calculate it.

Nevertheless, it can be estimated by using the survey data. The extent of the sampling error depends on many things, including the sampling method, the estimation method, the sample size and the variability of the estimated characteristic. This is why each sample estimate has its own sampling error. This error should thus be approximated for each estimated total, average, proportion, etc. produced by the survey.

Estimation of The Population Mean

In a simple random sample, the estimate of the population mean is identical to the mean of the sample:

$$\hat{x} = \frac{\sum x}{n}$$

where,

x = an observed value

$\hat{x}$ = estimate of the population mean

$\sum$ x = sum of all observed x values in the sample

n = number of observations in the sample.

Note: Lowercase x and n should be used if you are referring to a sample survey and upper case X and N should be used when referring to a population.

If the sample results have been summarized in a frequency table, then the estimate for the population mean is the same as the sample.

Thus,

$$\hat{x} = \frac{\sum xf}{\sum f}$$

where,

x = an observed value

f = the frequency of the value (the number of times that this value have been observed in the sample)

$\hat{x}$ = estimate of the population mean

$\sum xf$ = sum of all observed xf values (the product of the observed values times its frequency) in the sample

$\sum f$ = sum of the frequencies in the sample.

Example: A farmer randomly selects 10 eggs from a gross of 12 dozen eggs (144 eggs) he finds in his hen house. He carefully weighs each egg.

The following weights were recorded in grams: 0.75, 0.70, 0.55, 0.50, 0.60, 0.65, 0.75, 0.65, 0.75, 0.50

What is the mean weight of the gross of eggs?

Using the above formula, we can determine the mean weight of all of the eggs:

$$\hat{x} = \frac{\sum x}{n} = \frac{6.4}{10} = 0.64 \text{ grammes}$$

Estimation of The Population Total

For a simple random sample, the estimation formula of a total for the population is,

$$\hat{x} = N\frac{\sum x}{n}$$

where,

x = an observed value

$\hat{x}$ = estimated population total

$\sum x$ = sum of all observed x values in the sample

n = number of observations in the sample

N = total number of observations in the population.

It is just the estimate for the mean value multiplied by the number of units in the population.

In the previous example, the mean weight of an egg is 0.64 grams, so it is logical to think that the total weight of the 144 eggs would be 92.16 grams (144 × 0.64 = 92.16 grams). If sample results have been summarized in a frequency table, then the estimate formula for total population is,

$$\hat{x} = N\frac{\sum xf}{\sum f}$$

where

x = an observed value

$\hat{x}$ = estimated population total

$\sum xf$ = sum of all observed xf values in the sample

$\sum f$ = sum of frequencies in the sample

N = total number of observations in the population.

12

Measures of Dispersion

CENTRAL TENDENCY

The measures of central tendencies (*i.e.* means) indicate the general magnitude of the data and locate only the centre of a distribution of measures. They do not establish the degree of variability or the spread out or scatter of the individual items and their deviation from (or the difference with) the means.

- According to Nciswanger, "Two distributions of statistical data may be symmetrical and have common means, medians and modes and identical frequencies in the modal class. Yet with these points in common they may differ widely in the scatter or in their values about the measures of central tendencies."
- Simpson and Kafka said, "An average alone does not tell the full story. It is hardly fully representative of a mass, unless we know the manner in which the individual item. Scatter around it.... a further description of a series is necessary, if we are to gauge how representative the average is."

From this discussion we now focus our attention on the scatter or variability which is known as dispersion. Let us take the following three sets.

Students	Group X	Group Y	Group Z
1	50	45	30
2	50	50	45
3	50	55	75
mean ⇒	50	50	50

Thus, the three groups have same mean *i.e.* 50. In fact the median of group X and Y are also equal.

Now if one would say that the students from the three groups are of equal capabilities, it is totally a wrong conclusion then.

Close examination reveals that in group X students have equal marks as the mean, students from group Y are very close to the mean but in the third group Z, the marks are widely scattered. It is thus clear that the measures of the central tendency is alone not sufficient to describe the data.

Definition of dispersion: The arithmetic mean of the deviations of the values of the individual items from the measure of a particular central tendency used. Thus the 'dispersion' is also known as the "average of the second degree." Prof. Griffin and Dr. Bowley said the same about the dispersion.

In measuring dispersion, it is imperative to know the amount of variation (absolute measure) and the degree of variation (relative measure). In the former case we consider the range, mean deviation, standard deviation etc. In the latter case we consider the coefficient of range, the coefficient mean deviation, the coefficient of variation etc.

METHODS OF COMPUTING DISPERSION

- *Method of limits*:
  - The range
  - Inter-quatrile range
  - Percentile range
- *Method of Averages*:
  - Quartile deviation
  - Mean deviation
  - Standard Deviation and
  - Other measures.

Note that, we are going to study some of these and not all.

RANGE

In any statistical series, the difference between the largest and the smallest values is called as the range.

Thus Range (R) = $L–S$ $\begin{cases} L = \text{Largest value of the series.} \\ S = \text{Smallest value of the series.} \end{cases}$

Coefficient of Range: The relative measure of the range. It is used in the comparative study of the dispersion co-efficient of Range,

$$= \frac{L-S}{L+S}$$

Example: (Individual series) Find the range and the co-efficient of the range of the following items:

110, 117, 129, 197, 190, 100, 100, 178, 255, 790.

Solution: R = L–S = 790–100 = 690

$$\text{Co-efficient of Range} = \frac{L-S}{L+S} = \frac{790-100}{790+100} = \frac{690}{890} = 0.78$$

Example: (Continuous series) Find the range and its co-efficient from the following data.

Size:	10-20	20-30	30-40	40-50	50-10
Frequency:	2	3	5	4	2

Solution: R = L–S = 100–10 = 90

$$\text{Co-efficient of range} = \frac{L-S}{L+S} = \frac{100-10}{100+10} = \frac{90}{110} = 0.82$$

MEAN DEVIATION

Average deviations (mean deviation) is the average amount of variations (scatter) of the items in a distribution from either the mean or the median or the mode, ignoring the signs of these deviations by Clark and Senkade.

Individual Series

Steps:

- Find the mean or median or mode of the given series.
- Using and one of three, find the deviations (differences) of the items of the series from them.

  i.e. $x_i–x$, x_i–Me and x_i–Mo.

  Me = Median and Mo = Mode.
- Find the absolute values of these deviations *i.e.* ignore there positive (+) and negative (-) signs.

i.e. | *xi*–*x* |, | *xi*–Me | and *xi*–Mo |.

- Find the sum of these absolute deviations.

i.e. *S* | *xi*–*x* | +, *S* | *xi*–Me |, and *S* | *xi*–Mo |.

- Find the mean deviation using the following formula.

$$\text{M.D} = (\partial x) = \Sigma xi - x,\ \partial(\text{Me}) = \Sigma xi - \text{Me} \text{ or}$$
$$\partial(\text{Me}) = xi\ \text{Mo}$$

Note that:

- Generally M. D. obtained from the median is the best for the practical purpose.
- Co-efficient of M. D. = $\dfrac{\partial\overline{x}}{\overline{x}} = \dfrac{\partial(\text{Me})}{\text{Me}} = \dfrac{\partial(\text{Mo})}{\text{Mo}}$

Example: Calculate Mean deviation and its co-efficient for the following salaries: \$ 1030, \$ 500, \$ 680, \$ 1100, \$ 1080, \$ 1740. \$ 1050, \$ 1000, \$ 2000, \$ 2250, \$ 3500 and \$ 1030.

S.No.	**Salary (\$) xi**	**Deviation from Me (1080) ± ignored**
1	500	580
2	680	400
3	1030	50
4	1030	50
5	1050	50
6	1080	50
7	1100	30
8	1500	0
9	2000	420
11	2250	660
12	3500	920
		1170
		2420
n **= 13**		**Σ | *xi*–Me | = 6770**

Calculation:

- Median (Me)

$$= \text{size of } \left(\frac{n+1}{2}\right)^{\text{th}} \text{ item}$$

$$= \text{size of } 7^{\text{th}} \text{ item}$$

- $\text{M. D. } (\delta \text{ Me}) = \dfrac{\Sigma |x_i - \text{Me}}{n}$

$= \frac{6770}{13}$

= \$ 520.76

- Co-efficient of δ Me.

$= \frac{\delta Me}{Me} = \frac{520.76}{1080}$

Example: (Discrete series) Calculate the mean deviation of the following series.

Size:	4	6	8	10	12	14	16
Frequency	2	4	5	3	2	1	4

Solution:

Size(xi)	Frequency(f)	C.f.	| xi–Me |	| xi–Me |
4	2	2	| 8–4 | = 4	2 × 4 = 8
6	4	2 + 4 = 6	| 8–6 | = 2	4 × 2 = 8
8	5	6 + 5 = 11	| 8–8 | = 0	5 × 0 = 0
10	3	11 + 3 = 14	| 8–10 | = 2	3 × 2 = 6
12	2	14 + 2 = 16	| 8–12 | = 4	2 × 4 = 8
14	1	16 + 1 = 17	| 8–14 | = 4	1 × 6 = 6
16	4	17 + 4 = 21	| 8–16 | = 8	4 × 8 = 32
	n = 21			$\|x_i\text{–Me}\|$=68

Calculations:

- Median (Me) = Size of,

$$\left(\frac{n+1}{2}\right)^{\text{th}}$$

item = size of,

$$\left(\frac{21+1}{2}\right)^{\text{th}}$$

item

= Size of 11th item.

Therefore, Median (Me) = 8

- M. D. = $\frac{\Sigma|x_i - \text{Me}|}{\text{Me}} = \frac{68}{21} = 3.24$

Example: (Continuous series) Calculate the mean deviation and the coefficient of mean deviation from the following data using the mean. Difference in ages between boys and girls of a class.

Diff. in years:	No.of students:
0-5	449
5-10	705
10-15	507
15-20	281
20-25	109
25-30	52
30-35	16
35-40	4

Solution:

Diff.in age	Mid-values (x_i)	frequency (f_i)	$f_i x_i$	$\lvert x_i - \bar{x} \rvert$	$f_i \lvert x_i - \bar{x} \rvert$
0-5	2.5	449	1122.5	8	3592
5-10	7.5	705	5287.5	3	2115
10-15	12.5	507	6337.5	2	1014
15-20	17.5	281	4917.5	7	1967
20-25	22.5	109	2452.5	12	884
25-30	27.5	52	1430.0	17	884
30-35	32.5	16	520.0	22	352
35-40	37.5	4	150.0	27	108
		n=2123	$\Sigma f_i x_i$= 22217.5		$\Sigma f_i \lvert x_i - \bar{x} \rvert$ = 11440

Calculation:

1. $X = \dfrac{\Sigma f_i x_i}{n} = \dfrac{22217.5}{2123} = 10.5$ (approx.)

2. M. D. $= \dfrac{\Sigma f_i \lvert x_i - \bar{x} \rvert}{n} = \dfrac{11440}{2123} = 5.4$

3. co-efficient of M. D. $= \dfrac{\text{M.D.}}{\bar{x}} = \dfrac{5.4}{10.5} = 0.514$

VARIANCE

The term variance was used to describe the square of the standard deviation R.A. Fisher in 1913. The concept of variance is of great importance in advanced work where it is possible to split the total into several parts, each attributable to one of the factors causing variations in their original series.

Variance is defined as follows:

$$\text{Variance} = \frac{\Sigma(x_i - \bar{x})^2}{n}$$

Standard Deviation (s. d.)

It is the square root of the arithmetic mean of the square deviations of various values from their arithmetic mean. it is denoted by s.d. or σ. Thus, s.d.,

$$(\sigma x) = \sqrt{\frac{\Sigma(x_i - \bar{x})^2}{n}}$$

for the ungrouped data,

$$= \sqrt{\frac{\Sigma f_i(x_i - \bar{x})^2}{n}}$$

for the grouped data where $n = \Sigma f_i$

Merits

- It is rigidly defined and based on all observations.
- It is amenable to further algebraic treatment.
- It is not affected by sampling fluctuations.
- It is less erratic.

Demerits

- It is difficult to understand and calculate.
- It gives greater weight to extreme values.

Note that variance,

$$V(x) = \frac{\Sigma(x_i - \bar{x})^2}{n}$$

and *s. d.*,

$$(\sigma x) = \sqrt{\frac{\Sigma(x_i - \bar{x})^2}{n}} \text{ and } \sqrt{\frac{\Sigma f_i(x_i - \bar{x})^2}{n}}$$

Then, $V(x) = \sigma_x^2$

CO-EFFICIENT OF VARIATION (C. V.)

To compare the variations (dispersion) of two different series, relative measures of standard deviation must be calculated. This is known as co-efficient of variation or the co-efficient of s. d. Its formula is

$$\text{C. V.} = \frac{\sigma x}{\bar{x}} \times 100$$

Thus it is defined as the ratio s. d. to its mean.

Remark: It is given as a percentage and is used to compare the consistency or variability of two more series. The higher the C. V., the higher the variability and lower the C. V., the higher is the consistency of the data.

Example: Calculate the standard deviation and its co-efficient from the following data.

A	10	B	12	C	16	D	8
E	25	F	30	G	14	H	11
I	13	J	11				

Solution:

No No.	x_i	$(x_i - x)$	$(x_i - x)^2$
A	10	-5	25
B	12	-3	9
C	16	+1	1
D	8	-7	49
E	25	+10	100
F	30	+15	225
G	14	-1	1
H	11	-5	16
I	13	-2	4
J	11	-4	16
n = 10	**Σx_i = 150**		**$\Sigma \mid x_i - x \mid 2$ = 446**

Calculations:

- $\bar{x} = \frac{\Sigma x_i}{n} = \frac{150}{10} = 15$
- s.d.(σx) = $\sqrt{\frac{\Sigma(x_i - \bar{x})^2}{n}} = \sqrt{\frac{446}{10}} = 6.7$
- co-efficient of s.d. = $\frac{\sigma x}{\bar{x}} = \frac{6.7}{15} = 0.45$

Example: Prove that s.d. (sx) = $\sqrt{\frac{\Sigma xi^2}{n} - (\bar{x})^2}$

Solution: Now consider

$$\frac{\Sigma(x_i - \overline{x})^2}{n} = \frac{\Sigma(xi^2 - 2xi\overline{x} + \overline{x}^2)}{n}$$

$$= \frac{\Sigma xi^2}{n} - 2\overline{x}\frac{\Sigma xi}{n} + \overline{x}^2\frac{\Sigma 1}{n}$$

$$= \frac{\Sigma x_i^2}{n} - 2\overline{x}(\overline{x}) + \overline{x}^2.\frac{n}{n}$$

$$\ldots (\because \frac{\Sigma x_i}{n} = \overline{x} \text{ and } \Sigma 1 = n)$$

$$= \frac{\Sigma x_i^2}{n} - 2(\overline{x})^2 + (\overline{x})^2$$

Therefore, $\sigma x = \sqrt{\frac{\Sigma xi^2}{n} - (\overline{x})^2}$

This is known as a short-cut formula for standard deviation. The above problem can also be solved by this formula as:

Solution: We have $\Sigma x_i^2 = 2695$,
$n = 10$ and $\overline{x} = 15$ then

$$\text{s.d. } (s\ x) = \sqrt{\frac{2696}{10} - (15)^2}$$

$$= \sqrt{269.6 - 225} = \sqrt{44.6} = 6.67$$

Example: Calculate s.d. of the marks of 100 students.

Marks	No. of students(f_i)	Mid-values(x_i)	$f_i x_i$	$f_i x_i^2$
0-2	10	1	10	10
2-4	20	3	60	180
4-6	35	5	175	875
6-8	30	7	210	1470
8-10	5	9	45	405
	$n = 100$		$\Sigma f_i x_i = 500$	$\Sigma f_i x_i^2 = 2940$

Solution:

- $\overline{x} = \frac{\Sigma f_i x_i}{n} = \frac{5990}{1000} = 5.99$

$$\text{s.d.}(\sigma x) = \sqrt{\frac{\Sigma f_i x_i^2}{n} - (x)^2} = \sqrt{\frac{38770}{1000} - (5.99)^2}$$

- $\sqrt{38.77 - 35.88} = \sqrt{2.89} = 1.7$ (\$ in 100)

Example: Calculate s.d. of the marks of 100 students.

Marks	No. of students(f_i)	Mid-values(x_i)	$f_i x_i$	$f_i x_i^2$
0-2	10	1	10	10
2-4	20	3	60	180
4-6	35	5	175	875
6-8	30	7	210	1470
8-10	5	9	45	405
	$n = 100$		$\Sigma f_i x_i = 500$	$\Sigma f_i x_i^2 =$ 2940

Solution:

- $\frac{\Sigma f_i x_i}{n} = \frac{500}{1000} = 5$
- $\text{s.d.}(\sigma x) = \sqrt{\frac{\Sigma f_i x_i^2}{n} - (xy)^2}$

$$= \sqrt{\frac{2940}{100} - (5)^2} = \sqrt{29.40 - 25} = \sqrt{4.40}$$

Combined Standard deviation: If two sets containing n_1 and n_2 items having means x_1 and x_2 and standard deviations σ_1 and σ_2 respectively are taken together then,

- Mean of the combined data is $\bar{x} = \frac{n_1\bar{x}_1 + n_2\bar{x}_2}{n_1 + n_2}$
- s.d. of the combined set is

$$\sigma = \frac{n_1\left(\sigma_1^2 + d_1^2\right) + n_2\left(\sigma_2^2 + d_2^2\right)}{n_1 + n_2}$$

Where $d_1 = \bar{x}_1 - \bar{x}$ and $d_2 = \bar{x}_2 - \bar{x}$

Example: Adistribution consists of three parts, characterized as follows:

Part	No. of items	Arithmetic mean	Standard deviation
1	200	25	3
2	250	10	4
3	300	15	5

Also calculate its combined coefficient of variation.

Solution:

Part	Number	Mean s.d.
1	$n_1 = 200$	$\overline{x}_1 = 25\sigma_1 = 3$
2	$n_2 = 500$	$\overline{x}_2 = 10\sigma_2 = 4$
3	$n_3 = 300$	$\overline{x}_3 = 25\sigma_3 = 5$

- (Combined mean)123

$$= \frac{n_1\overline{x}_1 + n_2\overline{x}_2 + n_3\overline{x}_3}{n_1 + n_2}$$

$$= \frac{(23\times 200)+(10\times 250)+(15\times 300)}{200+250+300}$$

$$= \frac{12000}{750}$$

$$= 16$$

- $d_1 = \overline{x}_1 - \overline{x} = 25 - 16 = 9,$

  $d_2 = \overline{x}_2 - \overline{x} = 16 - 10 = 6$ and

  $d_3 = \overline{x}_3 - \overline{x} = 16 - 16 = 1$ and

  $d_1^2 = 81,\ d_2^2 = 36$ and $d_3^2 = 1$

- Combineds.d (σ)

$$= \sqrt{\frac{n_1\left(\sigma_1^2 + d_2^2\right) + n_2\left(\sigma_2^2 + d_2^2\right) + n_3\left(\sigma_3^2 + d_3^2\right)}{n_1 + n_2 + n_3}}$$

$$= \sqrt{\frac{200(9+81)+250(16+36)+300(25+1)}{200+250+300}}$$

$$= \sqrt{\frac{38880}{750}} = \sqrt{51.73} = 7.2$$

Combined coefficient of variation (C.V.)

$$= \frac{\sigma x}{\overline{x}}\times 100$$

$$= \frac{7.2}{16}\times 100 = 45\%$$

Example: The mean and s.d. of the marks of a group of 100 students in atest in Mercantile Law were given to be 40 and 3 respectively.

It was afterwards found that the marks of one student were wrongly recorded as 37 in stead of 73. Calculate the correct mean and s.d.

Solution:

- Incorrect mean = 40, n = 100

Using,

$$\bar{x} = \frac{\Sigma x_i}{n} \text{ we get}$$

Incorrect,

$$\Sigma x_i = n\bar{x} = 100 \times 40 = 4000$$

Therefore, correct mean,

$$(\bar{x}) = \frac{\text{Correct } \Sigma x_i}{n}$$

$$= \frac{4036}{100} = 40.36 \text{ marks}$$

- s. d. (sx) = $\sqrt{\frac{\Sigma x_i^2}{n} - (\bar{x})^2}$

Squaring we get $(\sigma x)^2 = \frac{\Sigma x_i^2}{n} - (\bar{x})^2$

Therefore, $\Sigma x_i^2 = n[(\sigma x)^2 + (\bar{x})^2]$

$$\begin{aligned}\text{Incorrect } \Sigma x_i^2 &= 100[(3)^2 + (40)^2] \\ &= 100[9 + 1600] \\ &= 160900 \\ \text{Correct } \Sigma x_i^2 &= \text{Incorrect } \Sigma x_i^2 - (37)^2 + (73)^2 \\ &= 160900 + 3960 \\ &= 164860\end{aligned}$$

Correct,

$$\sigma x = \sqrt{\frac{164860}{100} - (40.36)^2}$$

$$= \sqrt{19.6} = 4.43$$

Example: The score of two teams A and B in 10 matches are as:

A:	B:
40	21
32	14
0	29
40	13
30	5
7	12
13	10
25	13
14	30
5	0

Find the variance for both the series. Which team is more consistent ?

A			B		
x_i	$x_i - \bar{x}$	$(x_i - \bar{x})^2$	y_i	$(y_i - \bar{y})$	$(y_i - \bar{y})^2$
40	19	361	21	4	16
32	11	121	14	-3	9
0	-21	441	14	12	144
40	19	361	30	13	169
30	9	81	5	-12	144
7	-8	196	12	-5	25
13	-8	64	10	-7	49
25	4	16	13	-4	16
14	-7	49	30	13	169
9	-12	144	6	-11	121
$\Sigma x_i = 210$		$\Sigma(x_i - \bar{x})^2 = 1834$	$\Sigma x_i = 110$		$\Sigma(x_i - \bar{y})^2 = 862$

Example: Calculate the coefficient of variation of a series on the basis of the following results:

$n = 50, \Sigma(x_i - 75)$

$= -10, \Sigma(x_i - 7.5)2$

$= 400$

Solution:

- $\Sigma(x_i - 75) = \Sigma x_i - 375 = -10$... (since $\Sigma 7.5 = 7.5n = 7.5 \times 50 = 375$) $\Sigma x_i = -10 + 375 = 365$

  Therefore, $\bar{x} = \dfrac{\Sigma x_i}{n} = \dfrac{365}{50} = 7.3$

- $\Sigma(x_i-7.5)^2$
  $= \Sigma(x_i-7.3-0.2)^2$
  $= \Sigma(x_i-7.3-0.2)^2 - 0.4\ \Sigma(x_i-7.3) + 0.04_n$
  $= \Sigma(x_i - \bar{x})^2 - 0.4\Sigma(x_i - \bar{x}) + 0.04 \times 50$
  But $\Sigma(x_i - \bar{x}) = 0$
  $\Sigma(x_i-7.5)^2 = \Sigma(x_i - \bar{x})^2 - 0.4\Sigma\ (0) + 2 = 400$
  $\Sigma(x_i - \bar{x})^2 = 398$
  Now,

$$\sigma x = \sqrt{\frac{\Sigma(x_i - \bar{x})^2}{n}} = \sqrt{\frac{398}{50}}$$

  $= 2.8213$

- $\text{C. V.} = \frac{2.8213}{7.3} \times 100 = 38.65\%$

Solution:

- $\bar{x} = \frac{\Sigma xi}{n} = \frac{210}{10}$ and $\bar{y} = \frac{\Sigma yi}{n} = \frac{170}{10} = 17$

$$\sigma x = \sqrt{\frac{\Sigma(x_i - \bar{x})^2}{n}}\sqrt{\frac{1834}{10}} = \sqrt{183.4} = 13.54$$

Also $\sigma y = \sqrt{\frac{\Sigma(y_i - \bar{y})^2}{n}} = \sqrt{\frac{864}{10}} = \sqrt{86.2} = 9.28$

Therefore, $(\text{C.V})_A = \frac{13.54}{21} \times 100 = 64.47\%$ and,

$$(\text{C.V})_B = \frac{9.28}{17} \times 100 = 54.69\%$$

Since $(\text{C.V.})_A < (\text{C.V.})_B$ The base ball team A is more consistent.

Bibliography

Allan, C.: *Elements of Abstract Algebra,* New York: Wiley Publications, 2005.

Allen, T.: *Exploring Abstract Algebra with Mathematica,* New York: Pergamon Press, 2003.

Ayman, B.: *Abstract Algebra Manual: Problems and Solutions,* Carbondale: Southern Illinois University Press, 2003.

Celine, C.: *Abstract Algebra: Applications to Galois Theory, Algebraic,* New York: Holt, Rinehart, and Winston, 1998.

Claudia, M.: *Abstract algebra: A Comprehensive Treatment,* NY: Pergamon Publications, 1997.

Garrett, B.: *Selected Papers on Algebra and Topology,* Chicago: Nelson-Hall Publishers, 2003.

Israel K.: *A History of Abstract Algebra,* New York: Teachers College Press, 2007.

John R.: *Modern Algebra: An Introduction,* N.J.: Scarecrow Press, 2004.

Jonathan D.: *Introduction to Abstract Algebra,* IL: F.E. Peacock Publishers, 2008.

Joseph A.: *Contemporary Abstract Algebra,* New York: Falmer Press, 2005.

Nagpaul, S. R. *Topics in Applied Abstract Algebra,* Albany: State University of New York Press, 2004.

Richard E.: *Applications of Abstract Algebra with Maple,* New York: Plenum Press, 1999.

Robert, H.: *Abstract Algebra: A Concrete Introduction*, Ann Arbor: University of Michigan Press. 1999.

Theodore, S.: *Abstract Algebra: A Geometric Approach*, N.J.: Rowman & Littlefield, 1998.

Thomas, W.: *Abstract Algebra: Theory and Applications*, New York: Springer-Verlag, 2007.

William, P.: *Abstract Algebra: An Interactive Approach*, Washington: CQ Press, 2002.

Index